Soft Computing in Electromagnetics

Better communication systems demand high performance electromagnetic structures along with accurate, reliable and fast techniques to solve electromagnetic (EM) problems. A novel computing technique, called soft computing, is gaining popularity in a multitude of EM applications in order to tackle computationally intensive problems. It differs from conventional computing techniques by not relying on strict mathematical formulations. Soft computing techniques often seek to emulate biological systems like neural networks, swarm behaviour, etc. Fast-converging algorithms that mimic animal and human behaviour are currently emerging as the choice for replacing computationally intensive, time consuming, three-dimensional EM simulations; this development has simplified the process of EM design immensely.

Characterized by their ability to provide quick, robust and economically viable solutions despite imprecision, uncertainties and approximations in the formulation, soft computing methods such as genetic algorithm (GA), artificial neural network (ANN) and fuzzy logic have been widely used for microwave design. Similarly, they also play an important role in design and optimization applications in electromagnetics, such as EM design and performance enhancement of antennas, frequency selective surfaces (FSS), radar absorbing material (RAM) and metamaterials. This book emphasizes the suitability of soft computing techniques such as particle swarm optimization (PSO), bacterial foraging optimization (BFO) along with GA and ANN, for various EM design and optimization applications.

The application of soft computing concepts in the field of metamaterial antennas, radar absorbers, transmission line characterization and optimized radar absorbing material (RAM) is discussed in detail along with their usage for optimizing fault detection, EM propagation and path loss prediction. This book also introduces systematic implementation of soft computing tools in a relatively new area of metamaterials. Soft computing is presented here as an effective tool to minimize computations in a CAD package for quick and accurate solutions. The development of two such CAD packages for design of metamaterial split ring resonators (SRR) and path-loss prediction is presented. Numerical examples and MATLAB codes are provided to facilitate understanding of the principles of soft computing techniques by a wider readership.

Balamati Choudhury works as a Scientist at the Centre for Electromagnetics, CSIR-National Aerospace Laboratories, Bangalore. Her areas of interest include soft computing techniques, computational electromagnetics, and novel applications of metamaterials. She was recipient of the CSIR-NAL Young Scientist Award for the year 2013–2014 for her contribution in the area of Computational Electromagnetics for Aerospace Applications.

Rakesh Mohan Jha heads the Centre for Electromagnetics, CSIR-National Aerospace Laboratories, Bangalore. He worked as an SERC (UK) Post-Doctoral Research Fellow at Dept. of Engg. Sci., University of Oxford, England (in 1991–1992), and as Alexander von Humboldt Fellow at the Institute for High Frequency Techniques and Electronics of the University of Karlsruhe, Germany (in 1992–1993 and 2007). He was awarded Sir C.V. Raman Award for Aerospace Engineering for the Year 1999. Dr Jha was elected Fellow of INAE (FNAE) in 2010, for his contributions to the EM Applications to Aerospace Engineering.

Soft Computing in Electromagnetics

Methods and Applications

Balamati Choudhury

and

Rakesh Mohan Jha

CAMBRIDGE
UNIVERSITY PRESS

4843/24, 2nd Floor, Ansari Road, Daryaganj, Delhi - 110002, India

Cambridge University Press is part of the University of Cambridge.

It furthers the University's mission by disseminating knowledge in the pursuit of education, learning and research at the highest international levels of excellence.

www.cambridge.org
Information on this title: www.cambridge.org/9781107122482

First published 2015

Printed in India by Thomson Press India Ltd., New Delhi 110001

A catalogue record for this publication is available from the British Library

Library of Congress Cataloguing-in-Publication data

Choudhury, Balamati.
Soft computing in electromagnetics: methods and applications / Balamati Choudhury, Rakesh Mohan Jha.
    pages cm
Includes index.
Summary: "Discusses application of soft computing concepts in the field of metamaterial antennas, radar absorbers, transmission line characterization and optimised radar absorbing material (RAM)"-- Provided by publisher.
  ISBN 978-1-107-12248-2 (hardback)
1. Electromagnetic waves--Data processing. 2. Antenna radiation patterns--Data processing. 3. Radar--Data processing. 4. Electromagnetic devices--Materials--Data processing. 5. Absorption spectra--Data processing. 6. Soft computing. I. Jha, R. M. (Rakesh Mohan), 1959- II. Title.
TK7864.C44 2015
006.3--dc23
2015016642

ISBN 978-1-107-12248-2 Hardback

To

Professor Satya N. Atluri

Contents

List of Figures

List of Tables

Preface

At this point, we are at the throes of two revolutions — one is the information revolution and the other less visible one.... is the intelligent systems revolution.

—**Lofti Zadeh**

Ever since the days of Aristotle, classical scientific thinking has been based on strict logic, well-constructed definitions and mathematical expressions. This approach to science changed drastically when Dr Lofti Zadeh published his famous paper '*Fuzzy sets. Information and Control*' in 1965. By introducing imprecision in science, Dr Zadeh created in-roads into developing greater understanding in the field of artificial intelligence and even certain areas of philosophy and psychology! This imprecision, he claims, had led to a revolution in intelligent systems that has affected the way we live.

Today, the idea conceived by Dr Zadeh has grown into a whole new field of science—the field of soft-computing. Algorithms that attempt to mimic animal and human behaviour, evolution, etc., have been developed and implemented in problems ranging from scientific ones to even problems in economics and humanities! Certain researchers have also noted that soft computing techniques offer an alternate methodology to solve mathematically intensive problems.

The extension of this wondrous computation technique into one of sciences most mathematically challenging field, that of electromagnetics, is not surprising. This book address the implementation of soft computing in numerous, common electromagnetic problems. In doing so, computationally intensive, time consuming, three-dimensional electromagnetic simulations may be replaced by these fast-converging algorithms, thereby simplifying the process of electromagnetic design. This realization has led to a concerted effort by the Center for Electromagnetics, CEM (to which the authors are affiliated) towards improving existing research in soft computing. This book is a culmination of these efforts.

Accurate, reliable and fast optimization techniques are *a priori* requirements to cater to the demand for high performance, real time electromagnetic design objectives. Soft computing techniques are emerging as important tools in design and optimization of various complex electromagnetic problems. In view of this, an attempt has been made in this book to cover soft-computing based solutions to such EM problems. A brief overview of the topics covered in the book is given below.

Resolving problems such as fault detection and compensation in active antenna arrays are important for the aerospace community; finding out real time, cost effective solutions to these problems will help in handling critical situations. In addition, (i) need for miniaturized antennas, (ii) reduction of mutual coupling, and (iii) overall improvement in EM performance, are issues that concern antenna engineers worldwide. This book yields solutions to these issues through the soft-computing route, and gives a new perspective to solving such nonlinear problems.

This book also introduces the implementation of soft computing techniques in a relatively new area in science and technology—that of metamaterial and its applications. A user friendly CAD package for metamaterial *split ring resonator* (SRR) design using soft computing is also included in this book. Some of the important applications in electromagnetics such as antenna design and performance enhancement through *particle swarm optimization* (PSO) and bacterial foraging (BFO) have been included.

This book also covers the design and optimization of radar absorbing material (RAM) using PSO. The PSO algorithm is used to determine the optimum thickness of each layer of a Jaumann absorber followed by a more complicated problem statement, which necessitates the need for selection of materials from a database and optimizes the thickness of each layer of material for improved RAM performance. Later, the same algorithm is used to design metamaterial based RAM in both microwave and terahertz regimes.

Other topics covered in this book include the characterization of planar transmission line using artificial neural network (ANN) and a CAD package for ray-tracing in rural and urban environments.

To summarize, this book covers approaches to solving various complex electromagnetic problems through the novel route of soft computing. The theory behind these techniques is presented along with algorithms and the corresponding software codes. None of the books available so far covers such widespread topics and novel approaches towards real time and cost effective solutions.

Acknowledgments

At the outset, we wish to thank Mr Shyam Chetty, Director, CSIR-National Aerospace Laboratories, Bangalore for sustained support and official permission to write this book.

We would also like to acknowledge valuable suggestions from our colleagues at the Centre for Electromagnetics, Dr R.U. Nair, Dr Hema Singh, Dr Shiv Narayan and Mr K.S. Venu and their invaluable support during the course of writing this book.

It is our pleasure to acknowledge Ms Arya Menon who completed her off-campus undergraduate dissertation at CSIR-National Aerospace Laboratories. Indeed the work carried out by her was appreciated all around and we found it only fitting to invite her to adapt her dissertation as Chapter 8 of this book. We thank her for transferring the necessary copyrights for facilitating the production and circulation of the book in hand.

Beyond the technical aspects, correction of grammatical error is also a very pertinent area. We would like to extend our thanks to Mrs Balamani Vinayakumar for her help in going through the entire book for syntax error editing.

Without the concerted support of the publisher it is simply not feasible to write a book within a short span of time. Mr Manish Choudhary, Commissioning Editor, Cambridge University Press, has always been very responsive in this regard and we would like to thank him for all his inputs. We very much appreciate the meticulous effort made by Mr Hardip Grewal, Desk Editor (STM), Cambridge University Press, to improve the diction of the text and for ensuring that this book broadly conforms to the Cambridge University Press style and format.

Balamati Choudhury expresses her gratitude to her doctoral supervisors Prof. A. Patnaik (IIT Roorkee) and Prof. Ajit Panda (Dean, NIST, Berhampur) for their guidance and active help. She also wishes to thank her erstwhile NIST colleagues, Mr Om Prakash Acharya and Mrs Sandhya Pattnaik, for collaborative works in this exciting area of soft computing.

Needless to mention the gratitude, Balamati owes to her parents Mrs Binodini Choudhury and Mr Gajendra Kumar Choudhury and her uncle Mr Bipin Padhy for their blessings and constant encouragement. Further, she owes a lot to her brothers Balakrishna and Basudev, and their spouses Babita and Sujata, for their immense support coupled with a tinge of proud feeling of completion of this book.

R.M. Jha appreciates his wife Renu and daughter Vishnupriya for their support and putting up cheerfully with the demands on time that he had during the course of writing this book.

Abbreviations

AMC	Artificial magnetic conductors
ANN	Artificial neural network
BFO	Bacterial foraging
BGA	Binary genetic algorithm
BPSO	Binary particle swarm optimisation
BST	Barium strontium titanate
CD	Circular dichroism
CG	Conjugate gradient
CLPSO	Comprehensive learning particle swarm optimisation
CPGA	Continuous parameter genetic algorithm
CSRR	Circular split ring resonator
DLSR	Dual log-spiral resonator
DM	Dielectric materials
EBG	Electronic band gap
ECA	Equivalent circuit analysis
EM	Electromagnetic
ESS	Electromagnetic smart screen
FDTD	Finite difference time domain
FEL	Free electron laser
FEM	Finite element method
FSS	Frequency selective surface
GA	Genetic Algorithm
GNP	Gold nano-particles
HMM	Hyperbolic metamaterial
HZ-FSS	High impedance frequency selective surface
IPS	In-plane switching mode
IR	Infrared
LC	Liquid crystal
LDM	Lossy dielectric materials
LHM	Left-handed material
LIM	Low refractive index metamaterial

LMM	Lossy magnetic materials
MFDM	Multilayer finite-difference method
MIC	Microwave integrated circuits
MIMO	Multiple input, multiple output
MLP	Multi-layer perceptron
MLS	Method of least square
MOPSO	Multi-objective particle swarm optimisation
MOPSO	Multi-objective particle swarm optimisation
MTL-PSO	Multi-conductor transmission line particle swarm optimisation
NEP	Noise equivalent power
NN	Neural networks
NSGA	Non-dominated sorting genetic algorithm
PCS	Personal communication systems
PEC	Perfect electric conductor
PIFA	Planar inverted F antenna
PMM	Periodic method of moments
PRS	Partially reflecting surface
PSO	Particle swarm optimisation
RAM	Radar absorbing material
RCS	Radar cross section
RLM	Relaxation-type magnetic materials
RPSO	Real valued particle swarm optimisation
SLL	Sidelobe level
SRR	Split ring resonator
SSRR	Square split ring resonator
THz-TDS	Terahertz time domain spectroscopy
UWB	Ultra wideband
ZIM	Zero index metamaterial

Symbols

Lower case

a	Length of SRR
a_n	Amplitude distribution
c	Speed of light
c_1	Cognitive constant
c_2	Social constant
d	Spacing between array elements
d_z	Thickness of the metamaterial in the direction of wave propagation.
f	Transfer function
f_o	Centre frequency
f_{err}	Cost function for resonant frequency
f_m	Damping frequency
f_{mo}	Magnetic resonant frequency
f_r	Resonant frequency
g	gap between SRR ring
h	height of substrate
$\hat{i}_\theta$	Unit vector in the elevation direction
$\hat{i}_\phi$	Unit vector in the azimuth direction
i	number of input layer neurons
j	number of hidden layer neurons
k	number of output layer neurons
n	Refractive index
o	Output of the neural network
p	Solution search space
r_{ext}	External radius of SRR
s	Number of bacteria in search space
t	thickness
w	Width of SRR
w_{ik}	Weights of hidden layer
w_{eff}	Effective width of the strip
z	Impedance

Upper case

A	Amplitude
A_d	Amplitude of desired signal
AF_o	Instantaneous array factor
AFd	Measured array factor
A_{tar}	Total absorption
A_{iTM}	Absorption coefficient for TM polarization
A_{iTE}	Absorption coefficient for TE polarization
C	Gap capacitance
$Cpul$	per unit length capacitance
C_S	Effective capacitance
$C(i)$	Tumble step size in the random direction
E	Averaged squared error energy
E	Electric field
E_t	Tangential component of electric field
E_i	Incident field
E_T	Transmitted field
G	Antenna gain
H	Magnetic field
H_t	Tangential component of magnetic field
$J_{cc}(\theta, P(j, k, l))$	Cost function in BFO
$K(k)$	Complete elliptical integral
L	Total Inductance
M	Number of neurons
N	Number of antenna elements
N_p	Number of particles
N_d	Number of dimensions
N_t	Number of time steps
N_c	Number of chemotaxis steps
N_s	Number of swimming steps
N_{re}	Number of reproduction steps
N_{ed}	Number of elimination and dispersal steps
P_{ed}	Elimination-dispersal with probability
R	Reflectance
S_{11}	Scattering parameter from Port 1
S_{21}	Scattering parameter from Port 2
T	Transmittance
W	Weight matrix connecting the hidden to the output neurons
V	Weight matrix connecting the inputs to the hidden neurons
V_{min}	Minimum particle velocity
V_{max}	Maximum particle velocity
X_{min}	Minimum particle position
X_{max}	Maximum particle position
Y	Output from hidden layer neurons
Z_o	Impedance of free space

Greek

α	Attenuation constant
β	Progressive phase shift
δ	Intermediate error functions
ε	Permittivity of the medium
ε_o	Free space permittivity
ε_{eff}	Effective dielectric constant
ε_r	Relative permittivity
ε'_r	Real part of complex relative permittivity
ε''_r	Imaginary part of complex relative permittivity
η	Learning rate
η_o	Impedance of free space
θ	Elevation angle
λ	Wavelength
μ_o	Free space permeability
μ_r	Relative permeability
μ	Permeability of the medium
μ_i	Permeability of ith layer
μ'_{eff}	Real part of magnetic permeability
μ''_{eff}	Imaginary part of magnetic permeability
ρ	Filling factor of inductance
ϕ	Azimuth angle
ϕ_d	Azimuth angle of desired signal
ω	Angular frequency
Γ_0	Reflection coefficient

1 Introduction

Today's world can be best described by the term global village. Exploration, trade, commerce and scientific advancements have resulted in the breakdown of physical boundaries like mountains and oceans. While the credit for this virtual shrinking of the world is often attributed to the aforementioned factors, the biggest leap in this direction is a direct consequence of the discovery and subsequent improvements in communication systems—particularly wireless communication systems. The performance of these systems is an outcome of their constituent elements—elements whose designs are governed by electromagnetic (EM) formulations. Therefore, the contribution of electromagnetics in shaping the world as we currently know it, cannot be ignored. The demand for better communication systems has resulted in a demand for high performance electromagnetic structures, and as well as accurate, reliable and fast techniques to solve electromagnetic problems.

Almost in parallel with this demand for high performance in electromagnetics, a novel computing technique, called soft computing, is gaining popularity in a multitude of applications. This technique differs from conventional computing techniques by not relying on strict mathematical formulations. In fact, soft computing technique often seeks to emulate biological systems like neural networks, swarm behaviour, etc. Today, soft computing is increasingly being used to tackle non-linear, computationally intensive problems in engineering. Therefore, it is not surprising that these techniques find a comfortable niche in the field of electromagnetics, where there is a ubiquitous need for optimization.

1.1 Design and Optimization Scenarios

While the initial intention of soft computing was to address the problems in engineering design and optimization, the versatility of these techniques resulted in its application to almost all areas of day-to-day life including finance, humanities and medicine. A brief outlook on soft computing for these applications is discussed here.

1.1.1 Engineering applications

Soft computing plays an important role in providing cost effective and efficient means towards attaining the final objective of any engineering application, i.e., actual hardware realization. Hence, industrial applications of soft computing techniques in various fields such as microwave engineering, aerospace engineering, power systems, robotics, etc., are discussed here.

Design and optimization in microwave engineering applications is an important aspect that has been explored by various researchers. Soft computing techniques like *artificial neural network* (ANN) and *genetic algorithm* (GA) are proven to be effective design optimization tools in the field of microwave engineering [Patnaik and Mishra, 2000], and other soft computing techniques like *particle swarm optimization* [Robinson and Rahmat-Samii, 2004] and *bacterial foraging* [Datta and Misra, 2005] are emerging as fast optimization techniques [Choudhury *et al.*, 2012]. Computational time and accuracy are the two factors that need to be kept in mind before implementing any optimization technique. This book primarily deals with the implementation of all these techniques in electromagnetic applications with a focus on the key factors. Cost effective solutions are presented to common electromagnetic problems such as improvement of performance of antennas, mutual coupling reduction, fault detection in antenna arrays, radar absorbing material (RAM) design and optimization, metamaterial design and optimization, design and characteristics of transmission lines, and prediction of path loss in rural and urban environments.

Soft computing application holds its position in flight control as well as air traffic control systems [Napolitano *et al.*, 1999]. Spacecraft also implements soft computing for various applications such as docking operations, real time estimation of rover positions, etc., [Alvarez *et al.*, 1996]. Apart from flight control, other control systems in the circuit branches of engineering have explored the application of soft computing techniques in electric power systems, such as motion control of servo motors [Fahn *et al.*, 1999], plasma arc welding [Cook *et al.*, 1995], sensorless induction motor drives, etc., [Ben-Brahim *et al.*, 1999].

Advances in technology and a need for real-time human interaction with robots have necessitated development of computing systems that behave like human intelligence. Artificial neural network and fuzzy systems are the basic blocks for the development of these intelligent robotic systems. Mobile robots [Baranyi *et al.*, 2000], welfare robots [Takagi *et al.*, 1999], and emotional pet robots [Kubota *et al.*, 2000], are examples of systems that include neuro–fuzzy platform/ technology as the building blocks.

1.1.2 Medical applications

Applications of soft computing techniques are also well established in the field of medical science. These applications include medicine, diagnosis and surgical/ clinical applications. Yardimci [Yardimci, 2009] has discussed the research trend of soft computing in the field of medicine in the last decade. It has been observed that neural network has been used more effectively for search of medicine database compared to other techniques such as fuzzy systems, genetic algorithm, etc. Further a hybrid algorithm of fuzzy systems and neural network has been used in the science of clinical medicine. Shen *et al.*, [Shen *et al.*, 2005] used a hybrid neuro–fuzzy system for analysis of brain tissue from an MRI report.

1.1.3 Finance

Although soft computing was developed for engineering applications, it has also caught the attention of those in the financial trading sector. Specifically, these techniques have been used for the prediction of trend in future stock value [Tan, 2001; Quah and Srinivasan, 2000]. Soft computing techniques have been applied to a variety of markets such as Forex market [Chan and Foo, 1995], S&P 500 [Chenoweth et al.,1995], etc. The implementation of these techniques has been carried out at an academic level using neural network and fuzzy logic. The extension of these techniques into actual markets is still under process as developers need to accommodate practical trading constraints into the soft computing model [Vanstone and Tan, 2003].

1.1.4 Humanities and social sciences

The realm of soft computing has advanced to also include problems pertaining to humanities and social sciences. These techniques are mainly used in language processing and face recognition. Specifically, soft computing techniques are used to detect conditional and causal statements in language texts [Khoo et al., 1998], hypothesis and refutations, [Castiñeira et al., 2000], etc. The relationship between soft computing and the humanities is a two way relationship. On one hand, soft computing is used for the above mentioned applications; on the other hand, an effort is being made to model social behaviour into the development of soft computing based artificial intelligence models.

1.2 Electromagnetic Design Challenges

Electromagnetic design is a complicated procedure being dependent on the electrical length of the structure rather than its physical length. Further, material properties also contribute to the efficiency of the EM design. EM design involves three important steps, viz. processing, modelling and formulation. As design of electromagnetic structures is dependent on the frequency of operation and the material properties, care must be taken to choose the right material with the right electrical dimensions. As a result, performance improvements are often obtained through iterative manipulations of design geometry as well as material. Therefore, a compressive understanding of the issues faced during EM design will help the designer incorporate techniques to mitigate these issues during the design process. Ultimately, this technique will lead to quick, practical EM designs. In this section, sensitivity analysis, an important aspect of EM design towards actual hardware realization, has been taken up and discussed briefly.

Soft computing routes for EM designs often result in optimized design parameters for high performance. These optimized parameters may be accurate up to very high precision levels. However, the practicality of this accuracy, i.e., whether such designs may be achieved using the current fabrication technology, is a factor that must be discussed. For example, in adaptive antenna pattern synthesis problems, an optimized design may require a phase shifter of 2.5°, but only 6° phase shifters are commercially available. Similarly, an optimized metamaterial based absorber design may provide 99.9% absorption [Choudhury et al., 2013], whereas fabrication and material impurities may affect the performance of the EM design. Hence, it is always desirable to provide sensitivity analysis along with optimized design for practical applications.

1.2.1 Fabrication sensitivity

Fabrication issues depend on the type of fabrication technology as well as the EM design. Photo–lithography is a widely used fabrication technique for microwave designs. The photo–lithography equipment currently available can fabricate devices with minimum feature size of 2–2.5 μm. Hence sensitivity analysis in microwave range should be done with respect to the above mentioned size. Secondly, for design in the higher frequency ranges such as sub-millimetre wave and terahertz ranges, photo–lithography cannot be used for the fabrication of the optimized EM designs, as the design parameters may be smaller than 2 μm. However, these dimensions can be easily realized using electron–beam lithography, and this technique is recommended for fabrication of ultra-thin EM designs. However, a margin of 5–10% must be allowed for practical deviations from the design.

1.2.2 Material sensitivity

EM design and development of industrial hardware applications depend extensively on the intrinsic properties of materials. EM material parameters, such as permittivity and permeability of dielectrics, depend on the material synthesis and roughness. Similarly, reliability of the design depends on the accuracy to which the EM material parameters have been measured. Hence, considering the material issues, it is recommended to allow for a 5% shift in the material properties in order to compensate for any impurities in the material.

1.3   Objectives and Scopes

Soft computing techniques are emerging as important tools in design and optimization of various complex electromagnetic (EM) problems. In view of this, an attempt has been made in this book to cover the application of soft computing based solutions to EM problems.

Books that are currently available often focus on one particular algorithm of soft computing amongst *artificial neural network* (ANN), *particle swarm optimization* (PSO), *genetic algorithm* (GA), etc. No book examines the implementation of all these techniques for varied suitable applications. On the other hand, the proposed book examines the application of different soft computing techniques to various problems in electromagnetics.

Furthermore, another emerging algorithm of soft computing that is based on *bacterial foraging* (BFO) has also been introduced and implemented for the same class of problems along with a comparative study of computational efficiency and accuracy.

Some of the main objectives of this book are mentioned below:

- To provide low cost and real time solutions to various critical design problems. For example, resolving problems, such as fault detection and compensation in active antenna arrays, are important for the aerospace community. Finding out real time, cost effective solutions to these problems are demonstrated in this book using various soft computing techniques.

- To demonstrate solutions to non-linear problems concerning antenna engineers worldwide, such as (i) miniaturized antennas, (ii) reduction of mutual coupling, and (iii) overall improvement in EM performance.

- To introduce the implementation of soft computing techniques in a relatively new area in science and technology— metamaterial and its applications. A user-friendly CAD package for metamaterial *split ring resonator* (SRR) design using soft computing is also included here. Some of the important applications in electromagnetics such as antenna design and performance enhancement through PSO and BFO have also been incorporated.

- To describe the design and optimization of radar absorbing material (RAM) using PSO. The PSO algorithm is used to determine the optimum thickness of each layer of a Jaumann absorber followed by a more complicated problem statement, which necessitates the need for selection of materials from a database and optimization of the thickness of each layer of material for improved RAM performance. Later, the same algorithm is used to design metamaterial based RAM in both microwave and terahertz regimes.

- To help students and engineers learn implementation of soft computing techniques for their application domain effectively. In this regard, one dedicated chapter, which provides algorithms of soft computing techniques such as ANN, GA, PSO, BFO, etc., and implementation of these algorithms for solving simple mathematical function is discussed.

- To provide a comprehensive review of application of soft computing techniques in various state-of-the-art EM problems, for better exposure to the optimization scenario.

- Other topics covered in the book include the characterization of planar transmission line using artificial neural network (ANN) and a CAD package for ray-tracing in rural and urban environments.

To summarize, the main objective of this book is to provide effective solutions to critical EM design problems in a short frame of time through the soft computing optimization tool. The book provides information on the development of a computational engine that can integrate optimization techniques to EM solvers so as to achieve quick, highly efficient EM designs for non-linear applications where no analytic solution exists.

1.4   Organization of the Book

Soft computing methods play an important role in design and optimization in diverse engineering disciplines including those in electromagnetic (EM) applications. Soft computing techniques are characterized by their ability to provide quick, robust and economically viable solution(s) despite imprecision, uncertainties, and approximations in the formulation. The last decade has witnessed wider use of soft computing techniques such as *artificial neural network* (ANN), *fuzzy logic* and *genetic algorithm* (GA), in the RF and microwave domain. These include design and performance enhancement of antennas, *frequency selective surfaces* (FSS), *radar absorbing material* (RAM), metamaterial, etc. The aim of this book is to identify the suitability of other soft computing techniques such as *particle swarm optimization* (PSO), *bacterial foraging optimization* (BFO) along with *genetic algorithm* (GA) and *artificial neural network* (ANN) for various EM design and optimization applications.

The book also covers the use of soft computing techniques for some important electromagnetic applications such as fault detection in antenna arrays, path loss prediction in urban and rural areas and the design and optimization of metamaterials. Chapter 1 of the book provides an overview of the background of soft computing techniques in electromagnetics along with the challenges in implementation of these techniques.

Chapter 2 introduces the theoretical and algorithmic details of the techniques that come under the umbrella of soft computing. The focus is basically on evolutionary computing techniques like *genetic algorithm* (GA), *particle swarm optimization* (PSO) and *bacterial foraging optimization* (BFO) and the well-known *artificial neural networks* (ANN). These techniques can be used in all engineering applications including electromagnetics for the benefit of a wide range of readers. Therefore, in order to equip readers with a complete understanding of the concepts presented, an attempt has been made to describe the algorithms using specific examples. Computer codes provided at the end of each section in this chapter will also enhance the reader's understanding of the concepts.

The popularity and scope of soft computing in electromagnetics is brought out in Chapter 3. This chapter is an extensive literature review of the current trends in the field of soft computing in electromagnetics, which includes frequency selective surfaces, antenna miniaturization and performance enhancement, microwave devices, and invisibility cloaks.

Bacterial foraging optimization is an emerging soft computing technique. Literature on this algorithm is scarce. Therefore, in order to understand the behaviour of this algorithm as well as the effectiveness of its implementation in the field of electromagnetics, a chapter has been devoted to the implementation of BFO for practical problems. Chapter 4 discusses the design and performance enhancement of antennas using BFO. Towards this, BFO is implemented in order to optimize a metamaterial superstrate for a fractal antenna. The effect of the presence of this optimized layer on the gain and bandwidth of the antenna is discussed. Further, a similar optimization technique for mutual coupling reduction in elements of an antenna array is studied in detail. The ideas presented in this chapter are intended to facilitate the design of high performance, miniaturized antenna systems.

Absorbers find applications in various fields, the most important one being stealth technology. These absorbers, also called radar absorbers, are extremely complicated EM designs due to their inherent multilayer nature. Chapter 5 focuses on the implementation of PSO, a global optimization technique, for the optimization of radar absorbing material (RAM). A review of the application of various algorithms for optimization of conventional RAM is carried out, followed by a discussion on the PSO implementation of the same for Jaumann absorber and a multi-layered RAM. The technique is then extended to include optimization of metamaterial RAM designs in both the terahertz and microwave domains. While the RAM designed for microwave frequencies can be used in stealth technology, the terahertz RAM finds extensive application in terahertz biomedical spectroscopy.

Chapter 6 includes implementation of ANN for characterization of transmission lines. The analysis involves the determination of the characteristic impedance (Z_o) and the effective permittivity (ε_{eff}). The design includes finding of different dimensions of the transmission lines.

The properties of active antennas such as beam steering, null placing and radiation pattern synthesis has attracted the attention of the aerospace community. These active antennas feature a unique advantage of restoring the original radiation pattern through proper change of feeding distribution from the base-station. Chapter 7 focuses on the implementation of soft computing techniques for antenna array fault detection. In order to restore the original radiation pattern in a faulty antenna array, it is necessary to diagnose the array from the far-field perspective.

This problem has been solved through different routes using soft computing techniques such as ANN, GA, PSO, and BFO.

The chapters mentioned above discuss the application of single objective algorithms. Chapter 8 focuses on design of terahertz devices using multi-objective PSO. The algorithm involves the structural optimization of a terahertz metamaterial along with optimization of another application specific parameter.

EM designers prefer user friendly CAD models for analysis of various computationally intensive applications. Soft computing based CAD packages are gaining momentum in the field of electromagnetic (EM) applications because of their properties namely global optimization, quick response and accuracy. Chapter 9 discusses two in-house CAD packages developed using soft-computing techniques. The first CAD model uses PSO to design an optimized SRR structure for a specific frequency, while the second uses ANN, in order to predict path loss in rural and urban environments.

1.5   Summary

This book covers approaches to solving various complex electromagnetic problems through the novel route of soft computing. The theory behind these techniques is presented along with algorithms and their corresponding software codes. Emerging soft computing algorithms such as PSO and BFO are also tested for multi-objective optimization problems. Further, developments of computational engines that can integrate the optimization tool along with EM solvers are also discussed to provide a visibility into solving complicated non-linear problems, for which analytical formulations do not exist.

References

Baranyi, P., I. Nagy, P. Korondi, and H. Hashimoto, "General guiding model for mobile robots and its complexity reduced neuro-fuzzy approximation," *Proceedings of IEEE International Conference on Fuzzy Systems*, pp. 1029–1032, 2000.

Ben-Brahim, L., S. Tadakuma, and A. Akdag, "Speed control of induction motor without rotational transducers," *IEEE Transaction of Industrial Applications*, vol. 35, pp. 844–850, July/Aug. 1999.

Berenji, H. R., "Computational intelligence and soft computing for space applications," *IEEE Aerospace Electronic Systems Magazine*, vol. 11, no. 8, pp. 8–10, 1996.

Castineira E., E. Trillas, S. Cubillo, "On conjectures in ortho-complemented lattices," *Artificial Intelligence*, vol.117, pp. 255–275, 2000.

Chan, K. C. C. and K. T. Foo, "Enhancing technical analysis in the Forex market using neural networks," *Proceedings of IEEE International Conference on Neural Networks*, vol. 2, pp. 1023–1027, 1995.

Chenoweth, T., Z. Obradovic, and S. Lee, "Technical trading rules as a prior knowledge to a neural networks prediction system for the S&P 500 index," *Proceedings of IEEE Technical Applications Conference and Workshops*, pp. 111–115, 1995.

Choudhury, B., P. V. Reddy, S. Bisoyi, and R. M. Jha "Soft computing techniques for terahertz metamaterial RAM design for biomedical applications," *Computers, Materials and Continua*, vol. 37, no. 3, pp. 135–146, 2013.

Cook, G. E., R. J. Barnett, K. Anderson, and A. M. Strauss, "Weld modeling and control using artificial neural networks," *IEEE Transactions of Industrial Applications*, vol. 31, pp. 1484–1491, Dec. 1995.

Datta, T., and I. S. Misra, "A comparative study of optimization techniques in adaptive antenna array processing: The bacteria foraging algorithm and particle swarm optimization," *IEEE Antennas and Propagation Magazine*, vol. 51, no. 6, pp. 69–79, Dec. 2009.

Fahn, C. -S., K. -T. Lan, and Z. -B. Chen, "Fuzzy rules generation using new evolutionary algorithms combined with multilayer perceptrons," *IEEE Transactions of Industrial Electronics*, vol. 46, pp. 1103–1113, Dec. 1999.

Khoo, C. S. -G., J. Kornfilt, R. N. Oddy, S. H. Myaeng, "Automatic extraction of cause effect information from newspaper text without knowledge-based inferencing," *Literary and Linguistic Computing*, vol. 13, pp. 177–186, 1998.

Kubota, N., Y. Nojima, N. Baba, F. Kojima, and T. Fukuda, "Evolving pet robot with emotional model," *Proceedings of IEEE Conference on Evolutionary Computation*, pp. 1231–1237, 2000.

Napolitano, M. R., J. L. Casanova, D. A. Windon, II, B. Seanor, and D. Martinelli, "Neural and fuzzy reconstructors for the virtual flight data recorder," *IEEE Transactions of Aerospace Electronics Systems*, vol. 35, no. 1, pp. 61–71, 1999.

Patnaik, A., and R. K. Mishra, "ANN techniques in microwave engineering," *IEEE Microwave Magazine*, vol. 1, no. 1, pp. 55–60, 2000.

Quah, T. -S., and B. Srinivasan, "Utilizing neural networks in stock pickings," *Proceedings of the International Conference on Artificial Intelligence*, pp. 941–6, 2000.

Robinson, J., and Y. Rahmat-Samii, "Particle swarm optimization in electromagnetics, *IEEE Transactions on Antennas and Propagation*, vol. 52, no. 2, pp. 397–407, 2004.

Shen S., Sandham W, Granat M, Sterr A, "MRI fuzzy segmentation of brain tissue using neighbourhood attraction with neural-network optimization," *IEEE Transactions on Information Technology in Biomedicine*, vol. 9, pp. 459–467, 2005

Takagi, H., S. Kamohara, and T. Takeda, "Introduction of soft computing techniques to welfare devices," *Proceedings of IEEE Midnight-Sun Workshop on Soft Computing in Industrial Applications*, pp. 116–121, 1999.

Tan, C. N. W., *Artificial Neural Networks: Applications in Financial Distress Prediction and Foreign Exchange Trading*, Wilberto Publishing, Gold Coast, 113p., ISBN: 0957891105, 2001.

Vanstone, B., and C. Tan, "A survey of the application of soft computing to investment and financial trading," *Information Technology Papers*, pp. 211–216, Dec. 2003.

Yardimci A., "Soft computing in medicine," *Applied Soft Computing*, vol. 9, pp. 1029–1043, 2009.

Soft Computing Techniques

Soft computing is defined as a group of computational techniques based on artificial intelligence (human like decision) and natural selection that provides quick and cost effective solution to very complex problems for which analytical (hard computing) formulations do not exist. The term soft computing was coined by Zadeh [Zadeh, 1992]. Soft computing aims at finding precise approximation, which gives a robust, computationally efficient and cost effective solution saving the computational time. Most of these techniques are basically enthused on biological inspired phenomena and societal behavioural patterns. The advent of soft computing into the computing world was marked by research in machine learning, probabilistic reasoning, artificial neural networks (ANN), fuzzy logic [Jang *et al.*, 1997] and genetic algorithm (GA). Today, the purview of soft computing has been extended to include swarm intelligence and foraging behaviours of biological populations in algorithms like the particle swarm optimization (PSO) and bacterial foraging algorithm (BFO) [Holland, 1975; Kennedy and Eberhart, 1995; Passino, 2002].

Soft computing methods are associated with certain distinctive advantages. These include the following:

- Since Soft computing methods do not call for wide-ranging mathematical formulation pertaining to the problem, the need for explicit knowledge in a particular domain can be reduced.

- These tools can handle multiple variables simultaneously.

- For optimization problems, the solutions can be prevented from falling into local minima by using global optimization strategies.

- These techniques are mostly cost effective.

- Dependency on expensive traditional simulations packages can be reduced to some degree by efficient hybridization of soft computing methods.

- These methods are generally adaptive in nature and are scalable.

Of late, soft computing techniques have attracted recognition amongst researchers of various branches of engineering in order to arrive at solutions to problem statements [Patnaik and Mishra, 2000; Patnaik et al., 2005; Samii, 2006; Choudhury et al., 2012]. The sturdiness of the above techniques has been well tested pertaining to various problems encountered in every sphere of engineering. Indeed, the last decade has seen the implementation of soft computing in microwave applications. This chapter gives a glimpse of the various soft computing techniques that are widely used in the field of electromagnetics.

2.1   Artificial Neural Networks

Certain features of human brain such as the capability to recognize and perceive, have been studied for decades. The remarkable characteristics of the human brain drove researchers into attempting to emulate these characteristics in computers. Indeed, the outputs of these attempts have been incorporated into parallel processing systems and are collectively called as artificial neural networks (ANN). ANN, in the most general terms, is a network structure consisting of neurons (interconnected processing elements) with connection strengths (weights). They operate in parallel and the computation is performed at the processing element level [Haykins, 1999].

The artificial neural network was first developed by McCulloch in 1943. Some of the inherent capabilities of ANN are:

- Suitable for solving hard problems, as processing takes place in parallel instead of in serial mode [Christodoulou and Georgiopoulous, 2000].

- Capability to learn in both supervised and unsupervised mode. In supervised mode, the network is trained with the correct response, and in unsupervised mode, the network self-organizes and extracts patterns from the data presented to it [Wasserman, 1993].

- ANN is suitable for solving pattern matching problems [Haykins, 1999], pattern classification, optimization, self-organization, and associative memory.

- Hardware implementations of ANN systems are possible using VLSI technology [Zebulum et al., 2002].

2.1.1   Concept of ANN

An artificial neural network (ANN) is evolved from a biological neural network, which is made up of a large scale of neurons interconnected. From an engineering perspective, it can be regarded as an extension of conventional data–processing techniques. ANNs may be best perceived as a huge parallel processor, which is capable of storing knowledge gained through experience [c.f. Haykins, 1999]. The basic similarities between ANN and human intelligence are (1) ANN acquires the knowledge through learning, and (2) knowledge gained through experience is stored as connection strengths (synaptic weights).

From Fig. 2.1 it can be concluded that the basic unit of an artificial neural network is also a neuron, analogous to its biological counterpart. These neurons are capable of strengthening or weakening the connections between each other and are hence capable of adapting themselves to arrive at solutions, i.e. the neurons have the capability of adjusting interconnections to arrive at the desired output. An individual neuron does the same by adjusting its 'weights' and 'biases'.

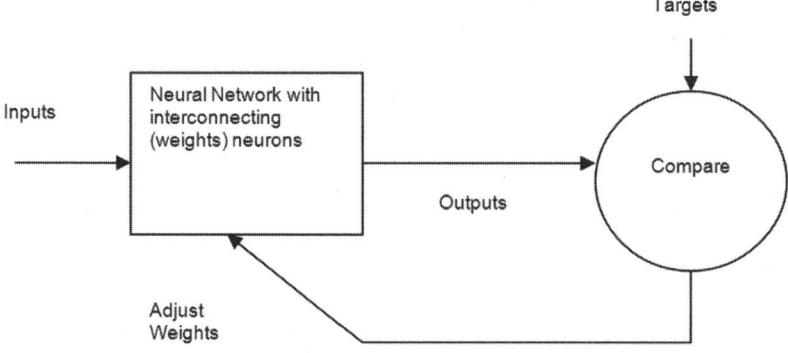

Fig 2.1 Basic concept of neural network

A neuron in an artificial neural network is modelled as shown in Fig. 2.2. Consider that a particular problem consists of N input variables. These variables can be represented as a matrix R of dimension $N \times 1$. As these variables are fed into the neuron, they are multiplied by weights w_i, where i ranges from 1 to N. The purpose of these weights is to manipulate the strength of the inputs. These weighted inputs are summed up and a bias, b is added. This bias is a scalar quantity and is capable of shifting the weighted sum. The output of these operations is denoted as x. The variable x is then used as the input to a transfer function. The output of the transfer function is considered to be the output of the neuron.

Therefore, the mathematical expression for the output of a neuron is given as

$$y = f(wR + b) \tag{2.1}$$

The transfer function, f, is often a standard function like a step function, linear function, log-sigmoidal function, etc., as shown in Figs. 2.3–2.5. However, a user defined transfer function can also be used. The choice of the transfer function is dependent on the problem in hand.

In its most basic form, a neural network consists of M_1 neurons placed in a parallel fashion. These M_1 neurons collectively represent a layer. The input R is fed to all the neurons in the layer. The output of a layer is hence a matrix of dimensions $M_1 \times 1$.

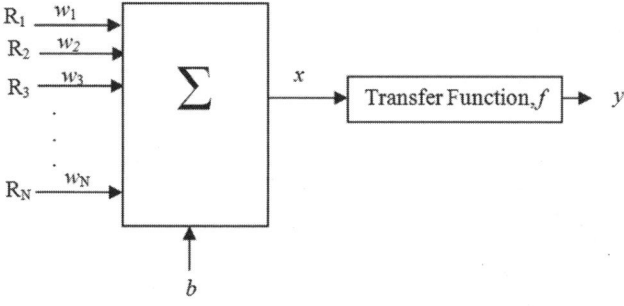

Fig 2.2 Structure of a neuron

As the complexity of the neural network increases, layers may be arranged one after the other with the output of one layer serving as the input to the other. The number of neurons in each layer need not be the same. For a neural network with T number of layers, the dimension of the output will be $M_T \times 1$, where M_T represents the number of neurons in the last layer. It must be noted that all intermediate layers are popularly called *hidden layers*. The implementation of an artificial neural network is a three–step process consisting of:

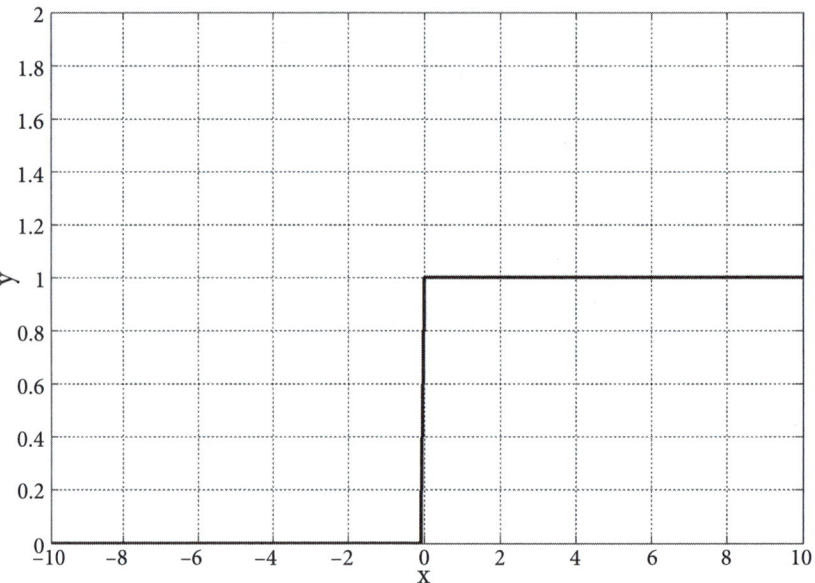

Fig 2.3   Step function: for x < 0, y = 0 and for x > 0, y = 1

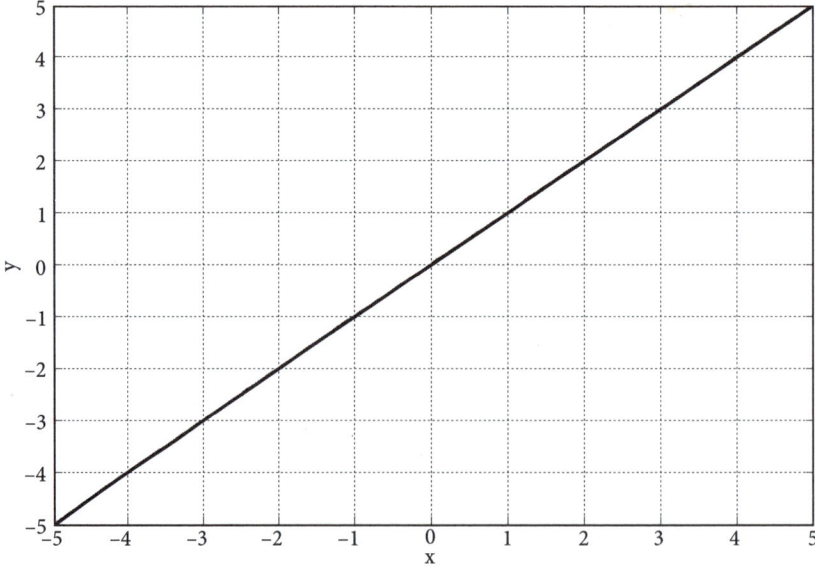

Fig 2.4   Linear function: x = y

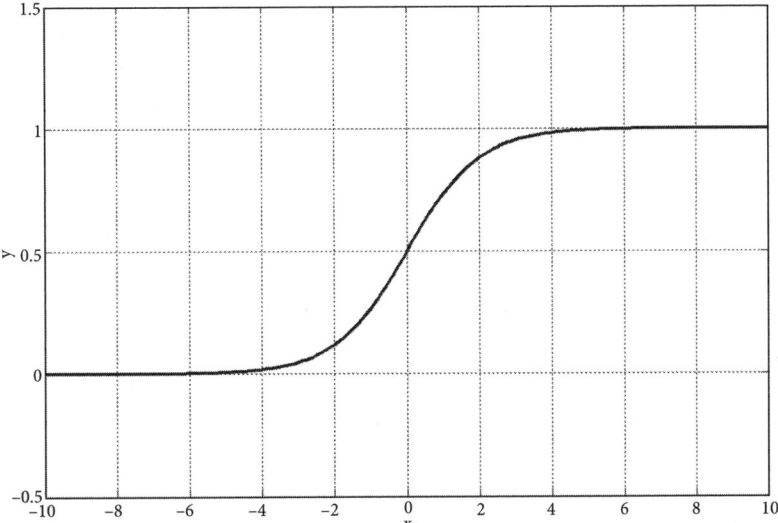

Fig 2.5 Sigmoidal function: for any value of x, the output, y, is restricted to values between 0 and 1

- *Data generation*:

  The functioning of a neural network relies heavily on the weights and biases of every neuron in all the layers. These values are determined by studying the input–output relationships in the given problem. As a result, a neural network requires a large set of accurate, pre-determined input and output data for the problem at hand. This data can be obtained from experimental observations or through simulation.

- *Training of network*:

  The neural network uses the available data in order to train the network for the input–output relationships. In essence, the neural network 'learns' to obtain the desired output by adjusting its weights and biases i.e. a neuron follows a certain learning rule, modifies its weight and bias, and trains itself to arrive at a certain class of outputs. This entire process is carried out using perceptrons—computer models that can simulate the ability of the brain to discriminate.

  The simplest perceptron, the single-layer perceptron, classifies linearly separable functions depending on defined thresholds. For a particular input, the perceptron calculates the output, compares the output to the threshold value, and classifies it. However, this model produces accurate results only for a single layer. In practice, multilayer perceptrons are used to represent non-linear separable functions where multiple hidden layers are involved. The multilayer perceptron uses a non-linear transfer function along with back propagation as the learning technique (discussed in the next section).

- *Testing*:

  After achieving the optimum weights and biases of the neurons, the neural network is tested for certain test vectors, and the outputs are observed and checked for accuracy. If the results obtained are of the desired accuracy, the values for the weights and biases can be considered to be finalized. The neural network can then be used without the generated data.

2.1.2 Back-propagation algorithm

In neural network theory, back-propagation is by far the most widely used algorithm used to solve regression problems [(Christodoulou and Georgiopoulous, 2000]. Widrow–Hoff learning rule has been extended to *multiple layer perceptron* (MLP) and nonlinear differentiable transfer functions. As defined by the user the target vector and the input vectors are used to train the neural network through an activation function and a set of hidden layer neurons. It is observed that a neural network having sigmoidal activation function with biases and linear output layer is capable of solving functions with a finite number of discontinuities.

The back-propagation algorithm is a gradient descent algorithm, where the term back-propagation refers to the approach for determining the gradient of nonlinear multilayer networks. In this algorithm, the network weights are stimulated with the negative of the gradient. Further, based on a few popular optimization techniques, such as Newton methods, conjugate gradient method, etc., a number of modifications in the basic algorithm have also been proposed. Properly trained back-propagation networks tend to give reasonable answers when new inputs are considered. Typically, a new input leads to an output closer to the accurate output. However, this works only if the new input is similar to the inputs used for training the neural network.

Flowchart and the step-by-step algorithm for implementation of the back-propagation algorithm are given in this section. The notations used in the description of this algorithm have the following meaning:

i : Neurons in the input layer

j : Neurons in the hidden layer

k : Neurons in the output layer

W: weight matrix connecting the hidden to the output neurons

V : weight matrix connecting the inputs to the hidden neurons

z : input to the network

y : output from hidden layer neurons

o : output of the network

Γ : transfer function

E : averaged squared error energy

d : desired output

δ : intermediate error functions

f : activation function

η : learning rate

For a specified training pairs P, $\{ z_1, d_1, z_2, d_2, ..., z_p, d_p \}$, where $i = 1, 2, ..., p$, d_i is $(k \times 1)$, and z_i is $(i \times 1)$. It should be noted that the ith component of each z_i takes a value -1 since input vectors have been augmented. Similarly, j^{th} component of y is of value -1; y is $(j \times 1)$ and o is $(k \times 1)$.

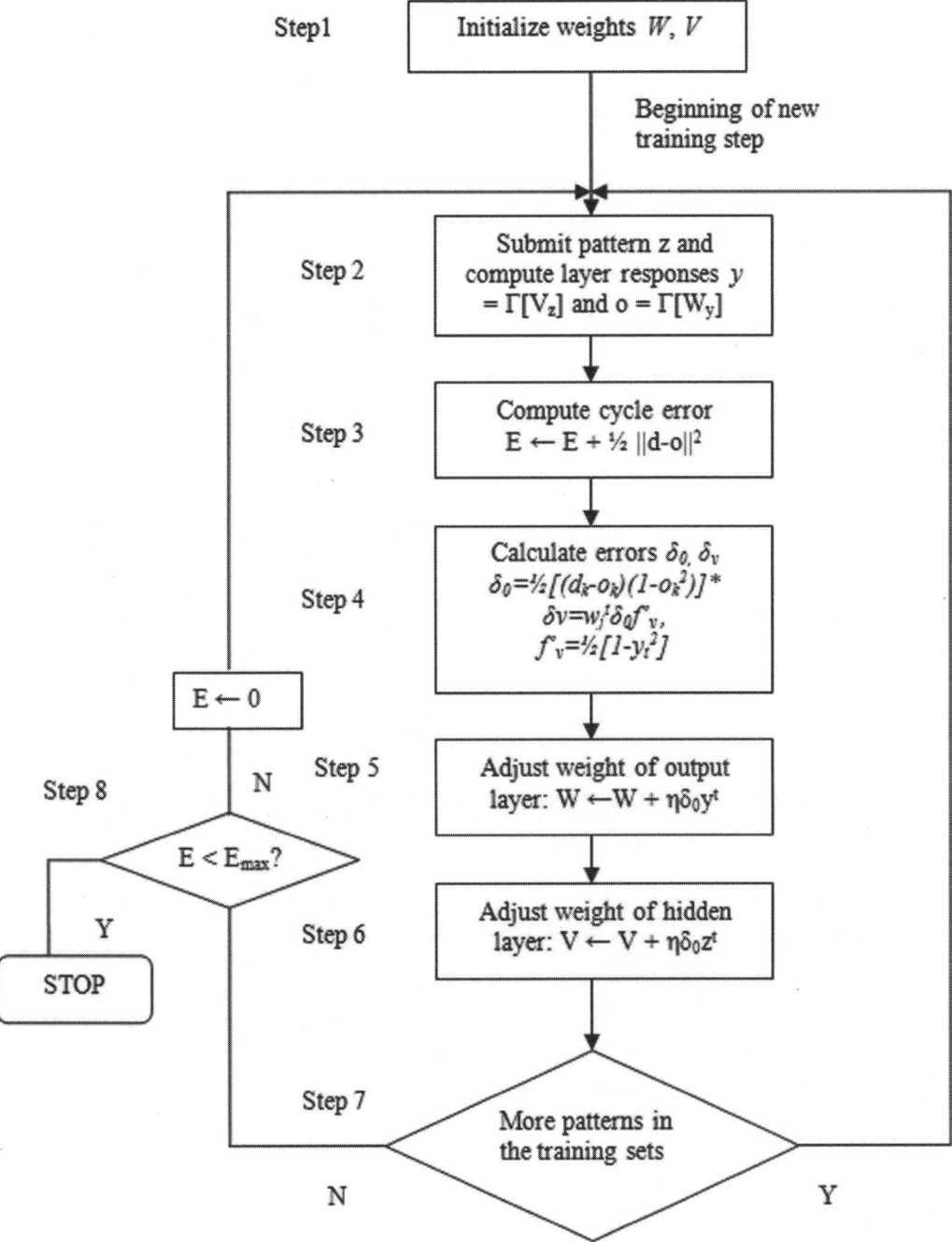

Fig 2.6 Error back-propagation training algorithm flowchart

Step 1: Choose learning rate $\eta > 0$, as E_{max}.

Initialize weights W and V as random values; where W is a matrix of $(K \times J)$, and V is a matrix of $(J \times I)$.

$$q \leftarrow 1, \, p \leftarrow 1, \, E \leftarrow 0$$

Step 2: This is the starting step of network training.

Compute outputs considering the inputs presented [f(net) is used]:

$$z \leftarrow z_p, \quad d \leftarrow d_p,$$

$$y_j \leftarrow f\left(v_j^t z\right)_, \qquad\qquad \text{for } j = 1, 2, ..., J$$

where v_j a column vector, is the kth row of V.

$$o_k \leftarrow f\left(w_k^t y\right) \qquad\qquad \text{for } k = 1, 2, ..., K$$

where w_k a column vector, is the kth row of W.

Step 3: The error value can be determined as:

$$E \leftarrow \frac{1}{2}\left(d_k - o_k\right)^2 + E, \qquad\qquad \text{for } k = 1, 2, ..., K$$

Step 4: For each layer the error signal vector δ_0 and δ_y are computed.

Where, vector δ_0 is $(K \times 1)$; δ_y is $(J \times 1)$.

In this step, the output layer error signal vector is

$$\delta_{ok} = \frac{1}{2}\left(d_k - o_k\right)\left(1 - o_k^2\right) \qquad\qquad \text{for } k = 1, 2, ..., K$$

In this step, the hidden layer error signal vector is

$$\delta_{yj} = \frac{1}{2}\left(1 - y_j^2\right)\sum \delta_{ok} w_{kj} \qquad\qquad \text{for } j = 1, 2, ..., J$$

Step 5: Weights of the output layer are attuned:

$$w_{kj} \leftarrow w_{kj} + \eta \delta_{ok} y_j \qquad\qquad \text{for } k = 1, 2, ..., ; j = 1, 2, ..., J$$

Step 6: Weights of the hidden layer are attuned:

$$v_{ji} \leftarrow v_{ji} + \eta \delta_{yj} z_i \qquad\qquad \text{for } j = 1, 2, ..., J; i = 1, 2, ..., I$$

Step 7: The loop conditions follow:

If $p < P$, then $p \leftarrow p + 1$, $q \leftarrow q + 1$, go to step 2, else go to Step 8.

Step 8: In this step the training cycle ends.

For $E < E_{max}$ the training session terminates.

Output weights W, V, q and E.

If $E > E_{max}$, then $E \leftarrow 0$, $p \leftarrow 1$: A new training cycle initiates through step 2.

The entire training process may be divided into the following four parts:

1. Training data generation
2. Creation of the neural network
3. Training of the created network
4. Simulation and testing of the trained network

2.1.3 Matlab code for ANN

The following code is an XOR classification problem. Four clusters of data (P, Q, R, S) are defined in a two dimensional space. The cluster pairs (P, R) and (Q, S) represents the XOR classification problem. The code given below uses the Matlab functions for training data and target classification.

```
close all;
clear all;
clc;
format compact;

% define number of samples of each class
N = 200;
% define 4 clusters of input data
del = .5; % offset of classes
P = [rand(1,N)-del; rand(1,N)+del];
Q = [rand(1,N)+del; rand(1,N)+del];
R = [rand(1,N)+del; rand(1,N)-del];
S = [rand(1,N)-del; rand(1,N)-del];
% Two classes are created. One with p and r and the other with
q and s
p = -1;
r = -1;
q = 1;
s = 1;

% The input to the neural network is obtained by combining the
data from the four clusters
X = [P Q R S];
```

```
% Target Y is defined
Y = [repmat(p,1,length(P)) repmat(q,1,length(Q))
repmat(r,1,length(R)) repmat(s,1,length(S)) ];
% Creation of neural network
net = feedforwardnet([5 3]);
% train net
net.divideParam.trainRatio = 1; % Set for training
net.divideParam.valRatio = 0; % Set for validation
net.divideParam.testRatio = 0; % Set for test
% train the neural network
[net,T,E] = train(net,X,Y);

% show network
view(net)
```

2.2   Genetic Algorithm (GA)

In this section, a brief overview of genetic algorithm along with terminologies are described, followed by the code for a simple program. *Genetic algorithm* (GA) uses the concept of natural selection and genetic inheritance [Goldberg, 1989]. The goals of GA are to understand and design system software based on natural selection. This algorithm maintains the robustness of natural systems. Genetic algorithm is a robust optimization procedure and the search algorithm is based on natural selection processes. The basic concept of this optimization is based on evolution, and the concept of the survival of the fittest.

2.2.1   Overview

Genetic algorithms, developed by John Holland in 1975, are motivated by Darwin's theory of natural evolution. According to the Darwinian Theory, in nature, the fit organisms will thrive whereas the unfit will not survive. Similarly, genetic algorithms search the space of individual organisms for better candidates. The chance of an individual being selected is proportional to the amount by which its fitness is greater or lesser than the competitor's fitness.

Genetic algorithms are different from normal search algorithms in the following manner:

- Coding of parameters (binary or real) is used instead of original values of the parameters.
- Search happens through a population of points, instead of a single point.
- It uses fitness function instead of derivatives.
- It uses probabilistic rules instead of the deterministic ones.

2.2.2   Terminologies of GA

As mentioned previously, the Darwinian theory of evolution proposes survival of the fittest. In effect, this ensures that the genes of the fittest organisms are passed on to the next generation while the genes of the weaker ones die off. Genetic algorithms draw a strong analogy to this transfer of genes. In essence, certain rules provide the foundation of genetic algorithms:

- Individuals in a population struggle for resources and mates; the one which is most suited to the environment often emerges as the winner.
- Individuals that are successful in each competition produce more offspring than those who are not successful.
- Genes from good individuals are passed on to the next generation. As a result, offspring of good parents are sometimes better than them.
- Each successive generation is given genes than make them more suited to the environment than its predecessor.

For a genetic algorithm, a population of individuals is maintained in a search space area. Each of these individuals represents a possible solution to the problem at hand. Finite vectors are used to code each individual. These vectors are composed of variables and generally, the variables belong to the binary set, i.e., each variable can take a value of either 0 or 1. These vectors are then compared to chromosomes—each chromosome (solution) consists of a finite number of genes (variables). The genes of each chromosome are evaluated and a fitness score is assigned to the chromosome. This value represents the individual's ability to compete in the population. By studying the genes in various chromosomes, the algorithm is able to combine information from them and generate offspring that are better suited to the environment than their parents. This technique is popularly called '*selective breeding*'.

The principle of the genetic algorithm is a static population size. Consider the initial size of the population to be n. The algorithm maintains the fitness associated with each individual in a database. Depending on this value, parents are selected to mate. Offspring are produced in accordance with a fixed reproductive plan. Due to this selection of parents, the offspring are often better than their parents. In addition, individuals with high fitness functions are given better opportunities to mate, thereby increasing the overall quality of the population. However, since the algorithm maintains a static size, individuals with low fitness values are eliminated in order to make space for better offspring. This process continues till the least fit solutions die out. The population eventually converges when each offspring is indistinguishable from the other. This value is then the solution to the given problem.

At the start of the algorithm, a random population is generated—this is the set of initial solutions. The algorithm then evolves through three operators [Goldberg, 1989]:

- *Selection/reproduction*: Equivalent to survival of the fittest
- *Crossover*: Mating between the pair of individuals of the population
- *Mutation*: Random modifications

2.2.2.1 Reproduction or selection

As described previously, a new population is generated by allowing the individuals with the best fitness to mate. As a result, the algorithm is fuelled by the hope that the new individual is significantly better than its parents. The fitness of each individual is obtained by evaluating an objective function using the chromosomes of the individual. The output of this fitness function determines the goodness of the individual, and ultimately if it should be allowed to mate or be exterminated. Encoding of the chromosome is mostly performed using binary strings. However, depending on the situation, other methods of encoding can also be followed.

2.2.2.2 Crossover

The crossover operator acts on certain selected genes from parent chromosomes yielding a new offspring. This operation makes GA easily distinguishable from other algorithms. After the selection operator chooses two individuals from the population, a crossover site along the string is randomly chosen. This represents the point after which the genes of one parent are replaced by the other. For example, consider two selected individuals with chromosomes as shown below.

Example: (| is the crossover point):

Chromosome 1	11011 | 00100110110
Chromosome 2	11011 | 11000011110
Offspring 1	11011 | 11000011110
Offspring 2	11011 | 00100110110

For a fixed crossover site, the selection of two chromosomes results in two offspring. The first possibility will have genes of Chromosome 1 before the crossover site and genes of Chromosome 2 after the crossover site. On the other hand, the other option will have genes of Chromosome 2 before the crossover site and genes of Chromosome 1 after the crossover site. Generally, both these offspring are added into the population of the next generation.

It should be mentioned here that the selection of the number of crossover points is application specific—certain applications can have more than one crossover point. These application specific crossovers improve the performance of GA.

2.2.2.3 Mutation

Life on earth is an outcome of natural selection. However, natural selection alone wouldn't have produced the variety of species that is seen today. In fact, had natural selection been the only evolutionary trend, life would have converged to a point where the environment was occupied by species highly adapted to it. The process of evolution, thus, would have slowed down many thousands of years ago and many brilliant life forms seen today wouldn't have even come into existence—perhaps even *homo sapiens*! Evidently, this does not occur in nature. Another phenomenon called mutation works in tandem with the process of natural selection. Mutation refers to the random changes in the genes of offspring. While some mutations result in desirable traits in offspring, others produce offspring that are less suited for the environment than their parents. When these mutated offspring mate, natural selection allows genes from the former to propagate, while the latter are exterminated.

Genetic algorithm also incorporates the phenomenon of mutation into its computation. The inclusion of mutation prevents premature convergence, i.e., a mutation ensures that the solutions don't fall into a local optimum of the problem. Further, the process of mutation also acts as a safeguard against irrevocable bit-flips [Goldberg, 1989] When genes are encoded in the binary form, mutation is achieved by randomly flipping (changing 1 to 0 and vice versa) a small portion of the net genes. This is illustrated in the example below:

Example: Original offspring 1   1101100100110110
               Original offspring 2   1101111000011110
               Mutated offspring 1   1101101100110100
               Mutated offspring 2   1100111000011110

It must be noted here that using mutation alone will result in random walk through the search space and the algorithm will never converge. In conclusion, by using both selection and mutation, the resultant algorithm is a parallel computing, noise-tolerant, hill climbing one.

Figure 2.7 gives the step-by-step procedure for implementation of the genetic algorithm. Let is assume that the algorithm is forced to stop when the number of generations reach a sufficiently large value N.

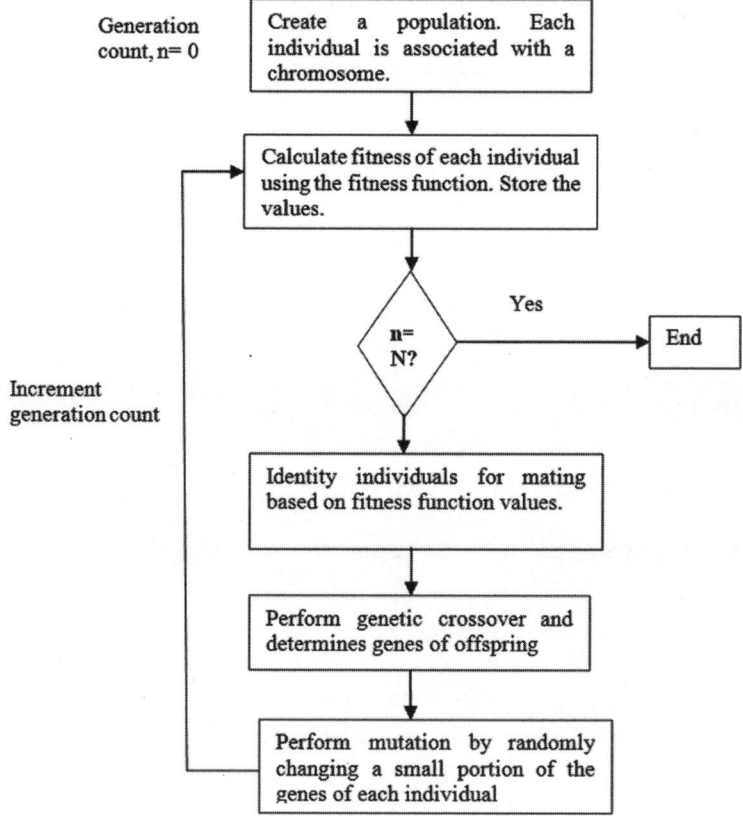

Fig 2.7   Flowchart for genetic algorithm

2.2.3 Matlab code for GA

The genetic algorithm is used to maximize the function $x^5 + x^2 + 1$. The value of x was allowed to range from [0, 31]. In the algorithm, each individual in the population could be represented as a

5 bit string [Goldberg, 1989]. As expected, at the end of algorithm, all the individuals converged to a value of 31. The code for the same is given below. Figure 2.8 shows the variation of the average fitness function for each generation. The run time of this code was found to be 0.2049 s.

```
%pop_real=value taken by individual (type: real)
%pop_str=value taken by individual (type: binary string)
%fitness=calculated fitness of each individual
%sumfitness=sum of fitness value of all individual
%Np=Number of individuals
%Nb=Length of binary string
%xMax=maximum value of pop_real
%xMin=minimum value of pop_real
%Nt=Total number of generations
%prob_mut=[probability of mutation, probability of no muta-
tion]
%-------------------------------------------------------------
tic
clc;
clear all;
Np=30;
Nb=5;
Nt=8;
xMax=31;
xMin=0;
prob_mut=[0.001
     1];
AverageFitnessHistory=[];

%Define random population over search space

%Initialize Particle Position
for p=1:Np
        pop_real(p,1)=xMin+(xMax-xMin)*rand;
        pop_real(p,1)=round(pop_real(p,1));    %Integer value
are required
end

%Convert Particles to binary strings
for p=1:Np
        pop_str=dec2bin(pop_real,Nb);
end
for p=1:Np
```

```
%Calculate the fitness for each individual
    fitness(p,1)=feval('func_eval_2', pop_real(p,:));
end

sumfitness=0;
%Find total sum of all fitness

for i=1:Np
    sumfitness=sumfitness+fitness(i,1);
end

%Start GA
for main_loop=1:Nt
%Begin reproduction
    mate=zeros(Np,1);
    for i=1:Np
        partsum=0;
        j=0;
        rand_val=rand*sumfitness;
    %wheel point calculator
        while (not((partsum>=rand_val)||(j==Np)))
            j=j+1;
            partsum=partsum+fitness(j,1);
        end
        mate(i,1)=(pop_real(j,1));
    end
    mate_str=dec2bin(mate,Nb);
    pop_str_old=pop_str;
%Store current population as old
    pop_real_old=pop_real;
    for p=1:Np
%Begin cross over and mutation

%Generate random point for cross-over
    cross_over_point=randint(1,1,[1,Nb-1]);
    for i=1:cross_over_point
%1st crossover solution
        pop1(1,i)=pop_str_old(1,i);
    end
    i=0;
    for i=cross_over_point+1:Nb
```

```
        pop1(1,i)=mate_str(1,i);
    end
    i=0;
    for i=1:cross_over_point
%2nd crossover solution
        pop2(1,i)=mate_str(1,i);
    end
    i=0;
    for i=cross_over_point+1:Nb
        pop2(1,i)=pop_str_old(1,i);
    end


%Perform mutation of both pop1 and pop2
    for i=1:Nb
        rand_val=rand;
        j=0;
        temp=0;
        while (not((temp>=rand_val)||(j==2)))
            j=j+1;
            temp=temp+prob_mut(j,1);
        end
        if j==1 %mutation happens for selected bit of pop1
            if pop1(1,i)==0
                pop1(1,i)=1;
            else
                pop1(1,i)=0;
            end
        end
        rand_val=rand;
        j=0;
        temp=0;
        while (not((temp>=rand_val)||(j==2)))
            j=j+1;
            temp=temp+prob_mut(j,1);
        end
        if j==1 %mutation happens for selected bit of pop2
            if pop2(1,i)==0
                pop2(1,i)=1;
            else
                pop2(1,i)=0;
            end
```

```
            end
        end

%select randomly
        sel_rand=randint(1,1,[1,2]);
        if sel_rand==1
          pop_str(p,:)=pop1;
        else
          pop_str(p,:)=pop2;
        end
        end

        pop_real=bin2dec(pop_str);
        for p=1:Np
%Calculate the fitness for each individual
        fitness(p,1)=feval('func_eval_2', pop_real(p,:));
end

%Find total sum of all fitness
sumfitness=0;
for i=1:Np
    sumfitness=sumfitness+fitness(i,1);
end
average_fitness=sumfitness/Np;
AverageFitnessHistory(main_loop)=average_fitness;

%Output the average fitness at every time step
        str = sprintf('Generation: %d Average Fitness: %f',
main_loop, average_fitness);
        disp(str)
end
display('Final population is')
display(pop_real);
figure(1)
    plot(AverageFitnessHistory);
    xlabel('Number of Iterations');
    hold on;
toc
function f=func_eval_2(x)
f=(x^5)+(x^2)+1;
end
```

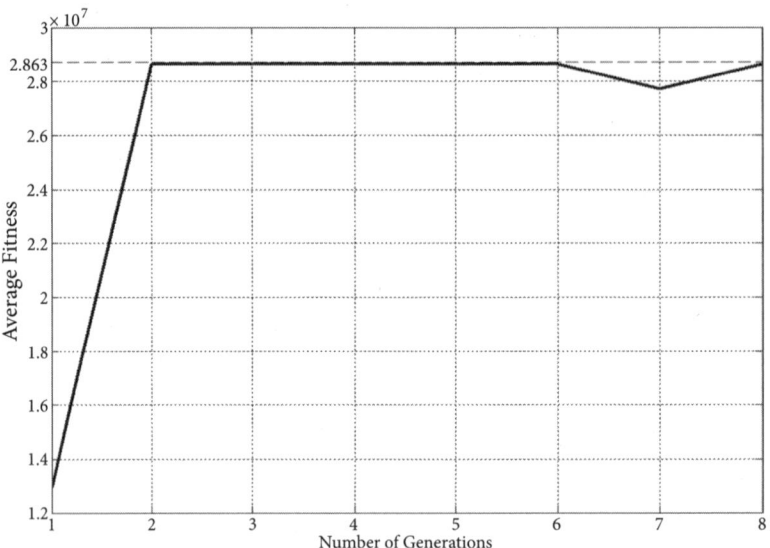

Fig 2.8 Variation of average fitness with respect to generations

2.3 Particle Swarm Optimization (PSO)

The dynamics of swarm behaviour is fascinating and awe-inspiring, and has driven many researchers to emulate this behaviour in fields such as robotics and computing. The most popular computing strategy based on swarm intelligence is *particle swarm optimization* (PSO). The strategy used by a swarm of bees for locating the highest density of flowers in a field (garden) is the inspiration for this algorithm. The bees start at a random location with a random velocity, and optimize the path to reach their goal without any prior knowledge of the field. This behaviour was first modelled by Kennedy and Eberhart [Kennedy and Eberhart, 1995] in order to create an optimizer.

2.3.1 Basic concept of PSO

The PSO is a multi-dimensional optimization technique based on swarm behaviour. This is a population based algorithm where each possible solution is referred as *particle*, represented by its position and velocity vectors. The problem space has been explored by the swarm of bees updating their position and velocity vectors to obtain the perfect solution. The particles continue with this updating process based on their own experience as well as that of the neighbourhood particles.

Let the number of decision parameters be D for a given optimization problem. Then the position of each particle in the swarm can be denoted as $X_i = (x_{i1}, x_{i2}, ..., x_{iD})$, and the velocity as $V_i = (v_{i1}, v_{i2}, ..., v_{iD})$. The algorithm updates this position and the velocity of each particle as:

$$v_{id} = v_{id} + c_1 r_1 (p_{id} - x_{id}) + c_2 r_2 (p_{gd} - x_{id})$$

(2.2)

$$x_{id} = x_{id} + v_{id} \qquad (2.3)$$

where, c_1 and c_2 are the cognitive constant and social constant, respectively, r_1 and r_2 are random numbers between [0, 1].

$P_i = (p_{i1}, p_{i2}, \ldots, p_{iD})$: Previous best position of each particle.

$P_g = (p_{g1}, p_{g2}, \ldots, p_{gD})$: Global best position of all particles.

The literatures reveal that the values of c_1 and c_2 should be equal (two positive constants). V_{max} is the maximum velocity vector on each dimension. A typical pseudo-code for classical *particle swarm optimization* is given below:

Step 1: Define the solution space.

Step 2: Define the position vector and velocity vector for all particles in the swarm of bees (randomly).

Step 3: Define fitness function f(X$_i$) and determine the value of fitness for each particle (personal best).

Check for condition: $f(X_i) < f(P_i)$ then $P_i = X_i$

Repeat Step 3 till condition satisfies.

Step 4: Update global best position P_g, if current $f(X) < f(P_g)$.

Step 5: Define r_1 and r_2 as random numbers.

Step 6: Compute particle velocity using Eq. 2.2.

Step 7: Repeat Step 3 to 6 until minimum error is achieved.

2.3.1.1 Binary PSO and real valued PSO (RPSO)

Real valued PSO is perhaps the most basic of all PSO algorithms. The algorithm operates in a real-valued solution space. Consequently, the swarm behaviour can be modelled using Newtonian mechanics and Eqs. 2.2 and 2.3 can be used directly in the algorithm.

However, at times, it is necessary to apply swarm behaviour to problems that require binary optimization. Let us assume that the dimension of the solution space is N. Each value in this space is a binary digit (0 or 1). Let the position of the mth agent at the tth iteration is given by

$$\vec{x}_{m,t} = \{x_{m1,t}, x_{m2,t}, \ldots \ldots, x_{mN,t}\} \qquad (2.4)$$

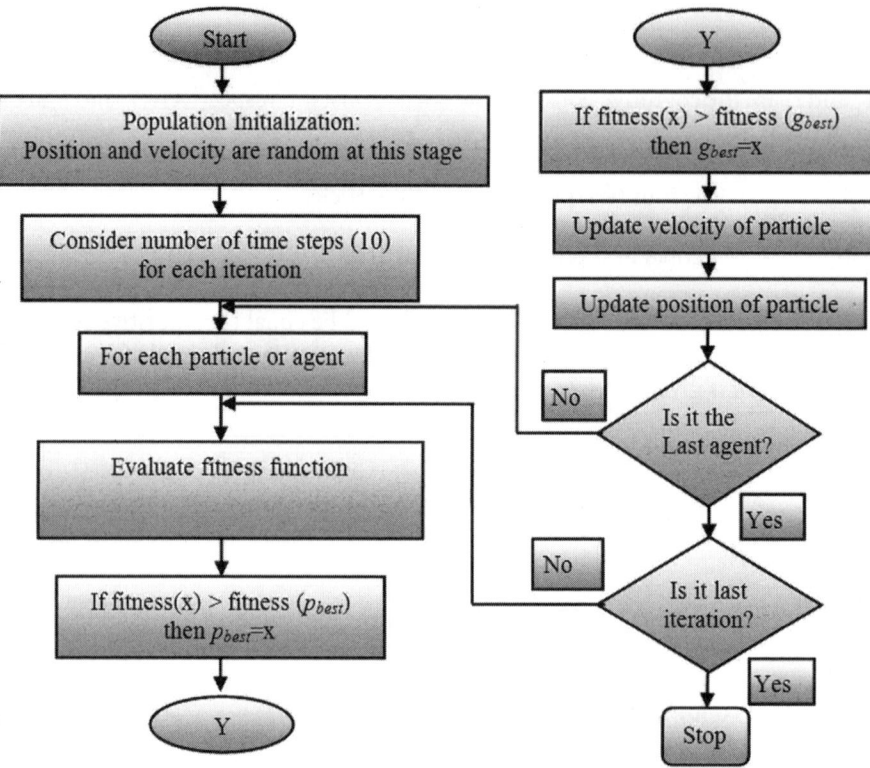

Fig 2.9 Flowchart of particle swarm optimization

If $p_{mn,t}$ is the personal best value of the agent at the tth iteration, then the velocity update depends on the subtraction of $x_{m,t}$ from $p_{mn,t}$. Since these values are binary strings, each element of the resultant string can take only three possible values: −1, 0 or 1. However, when multiplied by the Hooke's constants and added by momentum as given in Eq. 2.2, the resultant value of $x_{m,t+1}$ ceases to be binary and therefore does not lie in the solution space.

This dilemma was tackled by Kennedy and Eberhart in 1997 by introducing a transformation called *sigmoid limiting transformation*. This transformation was proposed to be an intermediate step in the velocity and position updates and is given by

$$S(v_{mn,t+1}) = \frac{1}{1+e^{-v_{mn,t+1}}}$$
(2.5)

Assuming that $v_{mn,t+1}$ lies within $[-V_{max}, V_{max}]$, and when $V_{max} \to \infty$, the resultant value of $S(v_{mn,t+1})$ lies with the range $(0,1)$. Now, a variable $r_{mn,t+1}$ with uniform distribution in $(0,1)$ is generated and the value for $x_{m,t+1}$ is decided such that

$$x_{mn,t+1} = \begin{cases} 1; r_{mn,t+1} < S(v_{mn,t+1}) \\ 0; r_{mn,t+1} > S(v_{mn,t+1}) \end{cases}$$
(2.6)

2.3.1.2 Single objective PSO and multi-objective PSO

When the objective of the PSO algorithm is fulfilled by a certain criteria (defined by a single fitness function), the algorithm is said to be a single object PSO. Depending on the nature of the solution space, the problem can be solved using either RPSO or BPSO as given in Section 2.3.2.1. However, some practical problems require the optimization of more than one fitness equation. Conventional PSO becomes inadequate in handling such problems and the usage of multi-objective PSO (MOPSO) comes into picture.

The idea is to assign a weight to each of these fitness parameters according to its order of importance in the design. Let us assume that K factors are considered for optimization. These correspond to K fitness functions. In order to convert this multi-objective design into a single objective one, each of these factors are weighted and consequently summed. Therefore, the single-objective equivalent of the multi-objective problem is given as

$$f(\vec{x}) = \sum_{i=1}^{K} w_i f_i(\vec{x})$$

(2.7)

The drawback of this method is the assigning of weights, since the magnitude of these weights is not known in many practical problems.

The most commonly used approach for solving multi-objective optimization problems involves the usage of the concept of Pareto dominance and was first proposed by Fieldsend and Singh in 2002. This approach is also known as directed MOPSO or d–MOPSO. The output of the d–MOPSO is a set of solutions that are non-dominated from each other. Depending on the problem, the user can choose any one of these solutions.

Pareto Front Multi-Objective PSO

As discussed previously, certain problem statements require optimization of more than one fitness function. In such scenarios, the aim of the PSO algorithm is to find points in the solution space wherein the fitness of one function can be improved only by degrading the fitness of the other functions. All such points are collectively called *pareto front*. Examples of some commonly obtained pareto fronts are given in Fig. 2.10. The solutions given by the pareto front are non-dominated with respect to each other and dominated with respect to all the other positions taken by the particles during the course of the simulation. In general, it is considered that *a* dominates *b* if [Jin and Samii, 2007]

$$f_i(a) \le f_i(b), \quad \text{and}$$

$$f_i(a) < f_i(b) \text{ for all } i$$

(2.8)

where, f represents fitness function.

The structure of the MOPSO algorithm is similar to that of RPSO discussed in the previous sections. However, unlike RPSO, MOPSO does not assign a single value for global best (*gbest*). Each particle in the solution space has its own *gbest*.

As the algorithm is run, all dominated solutions are added to a special matrix called '*archive*' [Jin and Samii, 2007]. During every step in the iteration, the archive is updated in order to consist of solutions that are non-dominated with respect to each other. The *gbest* of each particle is given by the nearest dominated solution (solution in archive) in terms of distance. For each particle, this *gbest* is used in the velocity update equation given in Eq. 2.2. The personal best of each particle (*pbest*) is the best dominated solution visited by the particle.

At the end of the iteration, the values in the archive represent the output of the MOPSO algorithm. These values represent points wherein the fitness of one function cannot be improved without degrading the fitness of the other functions. The user can choose any value from the archive depending upon the application and the importance assigned to each fitness function.

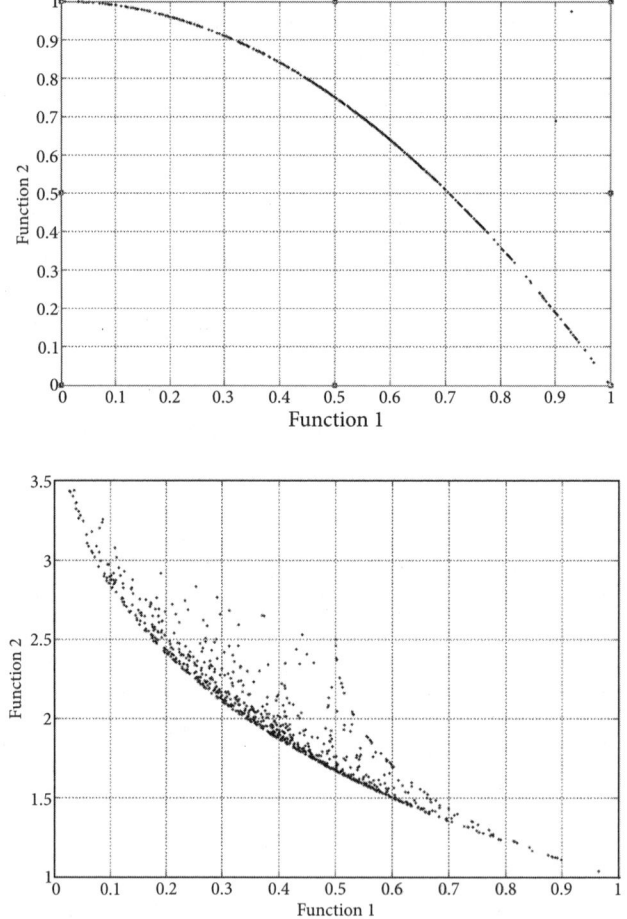

Fig 2.10 Common pareto-front geometries

Sigma PSO

Different techniques exist for finding the *gbest* for each particle [Mostaghim and Teich, 2003]; the technique mentioned in the previous section is only one of them. Another technique, called Sigma Method, can be used to determine the value of *gbest* for each particle. In this method, a value, σ is assigned to every particle in the solution space and in such a manner that all points on a radial line from the origin have the same value of σ [Mostaghim and Teich, 2003]. For a problem with two objective functions, this value is calculated using the following expression:

$$\sigma_i = \frac{f_{i,1} - f_{i,2}}{f_{i,1} + f_{i,2}} \tag{2.9}$$

where, f_1 and f_2 are the values of the fitness function for the ith particle. The value for σ is also computed for every entry in the archive. Finally, the *gbest* for a particle is that value in the archive for which the difference between σ of the particle and that of the archive entry is the smallest.

2.3.2 Matlab code of PSO

PSO code was written to find the optimum value for the following fitness function:

$$y = 1 - \frac{\sin\left[\pi\left(x_1 - 5\right)\right]}{\pi\left(x_1 - 5\right)} \frac{\sin\left[\pi\left(x_2 - 5\right)\right]}{\pi\left(x_2 - 5\right)}$$

```
clc;
clear all;
%-----------Define terms----------------------
w=0.25; %inertial Constant
c1=2;
c2=2;
Npar=20;
Ndim=2;
Nt=20;
xmin=0;
xmax=8;
vn=-1.6; %0.2*(xmax-xmin)
vx=1.6;
gbest_tracker = []; %Stores the value of gbest for every it-
eration

%Initialize Particle Position
R=zeros(Npar, Ndim);
for xp=1:Npar
    for i=1:Ndim
        R(xp,i)=xmin+(xmax-xmin)*rand;
    end
end
```

```
Vel=zeros(Npar, Ndim);
%Initialize particle velocity
for xp=1:Npar
    for i=1:Ndim
        Vel(xp,i)=vn+(vx-vn)*rand;
        if rand<0.5
            Vel(xp,i)=-Vel(xp,i);
        end
    end
end

%Evaluate fitness function at each point
M=zeros(Npar,1);
for xp=1:Npar
    M(xp,1)=feval('sincal', R(xp,:));
end
pbest_val=M;
%Declare Initial Personal Best
pbest_pos=R;
[gbest_val, index]=min(pbest_val);
%Declare Initial Global Best
gbest_pos=R(index,1:Ndim);

for j=1:Nt
%Begin Optimization Algorithm
    for xp=1:Npar
%Update Velocity
        for i=1:Ndim
            Vel(xp,i)=(w*Vel(xp,i))+(c1*(pbest_pos(xp,i)-
    R(xp,i))*rand)+(c2*(gbest_pos(1,i)-R(xp,i))*rand);
            if Vel(xp,i)>vx
                Vel(xp,i)=vx;
            elseif Vel(xp,i)<vn
                Vel(xp,i)=vn;
            end
        end
    end

    for xp=1:Npar
%Update position
        for i=1:Ndim
```

```
        R(xp,i)=R(xp,i)+Vel(xp,i);
            if R(xp,i)>xmax
%correct errors
                R(xp,i)=xmax;
            elseif R(xp,i)<xmin
                R(xp,1)=xmin;
            end
        end
    end

    for xp=1:Npar

%Evaluate Fitness

        M(xp,1)=feval('sincal', R(xp,:));
        if M(xp,1)< pbest_val(xp)

%Check if personal best
            pbest_val(xp)=M(xp,1);
            pbest_pos(xp,:)=R(xp,:);
        end
    end

%Determine gBest value and Position
    [cont_val, cont_pos]=min(pbest_val);
    if cont_val<gbest_val
        gbest_val=cont_val;
        gbest_pos=pbest_pos(cont_pos,:);
    end
    figure(1)
    plot(R(:,1),R(:,2),'x');
    axis([-0 8 0 8]);
    pause(0.5);
    hold on;
    gbest_tracker (j)=gbest_val; %Store the value of gbest
value
    gval = sprintf('Global best in iteration %d is %f', j,
gbest_val);
    disp(gval)
end
```

```
disp('Position')
disp(gbest_pos)
    figure(2)

%Plot history of best fitness
    plot(gbest_tracker);
    xlabel('Number of Iterations');
    ylabel('Value of fitness Funtion');

%Define sinc function (acts as fitness function)
function op=sincal(par_ip)
ff1=sin(pi*(par_ip(1)-5))./(pi*(par_ip(1)-5));
ff2=sin(pi*(par_ip(2)-5))./(pi*(par_ip(2)-5));
op = 1-(ff1.*ff2);
end
```

The best fitness was obtained for $x_1 = x_2 = 5$. The variation of the fitness function with respect to number of iterations is given in Fig. 2.11. Figure 2.12 shows all the values taken by the particles during the entire run of the algorithm. It can be seen that the particle density is highest near the co-ordinate (5, 5) as the solution is present at this point.

2.4   Bacterial Foraging Optimization

As compared to the PSO, the *bacterial foraging optimization* (BFO) is relatively new in the field of electromagnetics. This technique was introduced by K. M. Passino in 2002

2.4.1 Basic concept

Natural selection is the basic concept of *bacterial foraging optimization,* where the animals with poor foraging behaviour are expelled and the genes of the fittest animals with successful foraging capability are transmitted to the next generation. The bacterial foraging algorithm is an evolutionary computational algorithm developed on the basis of foraging behaviour of *E. coli* bacteria. *E. coli* is a bacterium, which lives in the gut, has 1 μm diameter and 2 μm length, and reproduction efficiency at a rate of 10–7 by means of splitting in 20 min. *E. coli* has the ability to move up to 100–200 rps by means of six rigid flagella.

2.4.2 Terminologies in BFO

The terminologies of BFO follow the foraging strategies of *E. coli* bacteria namely *chemotaxis, swarming, reproduction,* and *elimination and dispersal.* The details about the same are given below.

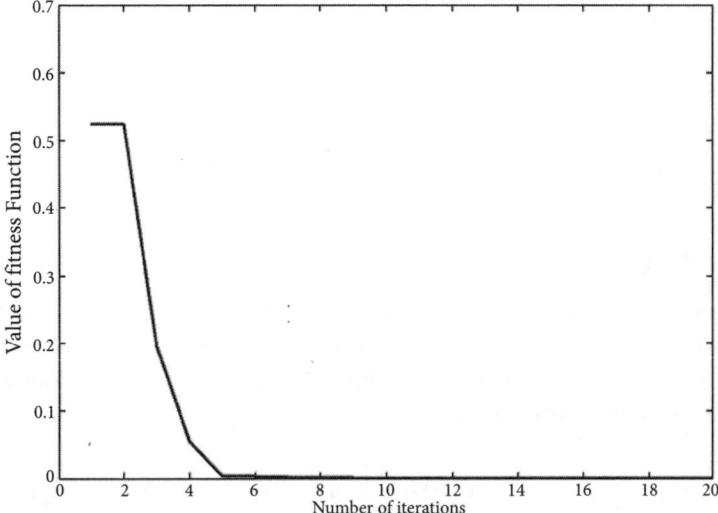

Fig 2.11 Variation of fitness function

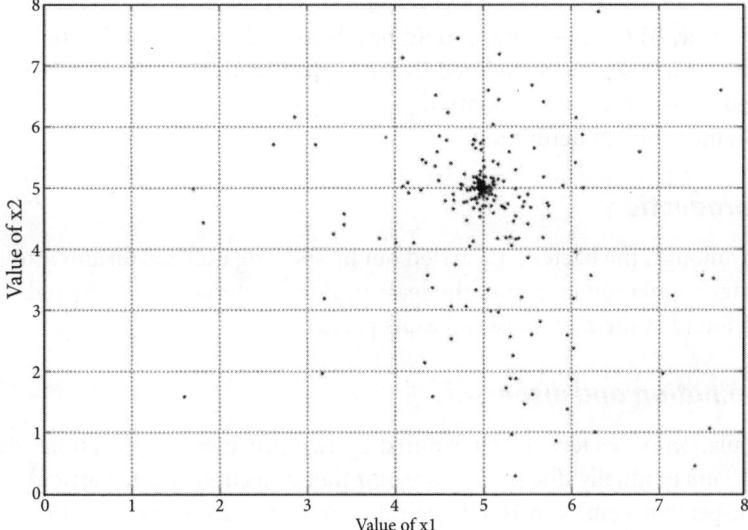

Fig 2.12 Position of particles for all PSO iterations

2.4.2.1 Chemotaxis

This process defines movement by flagella in a particular direction (*swimming* and *tumbling*).
The rotation of flagella in each bacterium determines the way in which it moves. The movement
in a predefined direction is called swimming and in different directions, is called as tumbling.
Let us consider $\varphi(v)$ as a unit length random direction, which represents a tumble. After each
tumble this $\varphi(v)$ will define the direction of movement.

In particular,

$$\theta^u(v+1, w, l) = \theta^u(v, w, l) + C(u)\varphi(v) \tag{2.10}$$

where, $\theta^u(v, w, l)$ represents the uth bacterium at v^{th} chemotactic, wth reproductive and l^{th} elimination and dispersal step. $C(u)$ is the tumble step size in the random direction.

2.4.2.2 Swarming

During the course of getting towards the best site for food, the bacterium which is seeking the best path, should try to attract other bacterium also to swarm together and reach the destination. The bacterial come together into groups with high bacterial density. Mathematically the swarming process can be represented as:

$$J_{cc}\left(\theta, P(v,w,l)\right) = \sum_{u=1}^{N} J_{cc}^u\left(\theta, \theta^u(v,w,l)\right) = \sum_{u=1}^{N}\left[-d_{attract}\exp\left(-\omega_{attract}\sum_{m=1}^{q}\left(\theta_m - \theta_m^u\right)^2\right)\right]$$
$$+ \sum_{u=1}^{N}\left[h_{repellent}\exp\left(-\omega_{repellent}\sum_{m=1}^{q}\left(\theta_m - \theta_m^u\right)^2\right)\right] \tag{2.11}$$

where, $J_{cc}(\theta, P(v, w, l))$ the value that should be added to the cost function to be optimized. N is the total number of bacteria considered in the algorithm and q is the number of parameters to be optimized. Further, the coefficients, $d_{attract}$, $\omega_{attract}$, $h_{repellent}$, $\omega_{repellent}$ are to be chosen properly and depend on the problem definition.

2.4.2.3 Reproduction

Reproduction amongst the bacteria is carried out by splitting each bacterium into two. In order to keep the total population constant, the least healthy bacteria are eliminated. The daughter organisms are found at the same position as its parent.

2.4.2.4 Elimination and dispersal

The local population of bacteria is determined by two processes: elimination and dispersion. Elimination occurs gradually due to the consumption of nutrients in a particular area. On the other hand, dispersion results in the dispersal of bacteria from one area into another. This event occurs due to some local environmental phenomena. Both these processes destroy the progress of the chemotatic step. However, it must be pointed out that dispersal often results in the placement of bacteria near a food source—a highly beneficial situation. Overall, these two events prevent the algorithm from falling into a local optima, thereby preventing stagnation.

2.4.3 Algorithm of BFO

The algorithm of BFO is as follows [K. M. Passino, 2005]

Step 1: Define and initialize parameters of the bacterial foraging algorithm with respect to the problem definition such as search space dimension (p), number of bacteria (S), number of chemotaxis steps (N_c), number of swimming steps (N_s), size of the step to considered for tumble $(C\ (u)$. where $u\ =\ 1,2,.....S)$, number of reproduction steps (N_{re}) and number of elimination and dispersal steps (N_{ed}) and probability of elimination-dispersal (P_{ed}).

Step 2: Increment the elimination-dispersal loop by one, i.e., $l = l + 1$

Step 3: Increment the reproduction loop by one, i.e., $w = w + 1$

Step 4: Increment the chemotaxis loop by one, i.e., $v = v + 1$

- For each bacterium $u = 1, 2,....S$, determine the cost function $J\ (u,\ v,\ w,\ l)$ using Eq. 2.11 as:

$$J\ (u,\ v,\ w,\ l) = J\ (u,\ v,\ w,\ l) + J_{cc}\ (\theta^{u}\ (v,\ w,\ l),\ P\ (v,\ w,\ l))$$

- Save the value of $J\ (u,\ v,\ w,\ l)$ as J_{last} to check for further improvement in the result.

- Generate random vector for tumble as: $\Delta(u) \in \Re^{p}$

where, $\Delta_{m}(u)$, $m = 1, 2, ..., p$. Define the step size $C\ (u)$ in the direction of the tumble and compute

$$\theta^{u}(v+1,w,l) = \theta^{u}(v,w,l) + C(u)\frac{\Delta(u)}{\sqrt{\Delta^{T}(u)\Delta(u)}}$$

- Determine the value of $J\ (u,\ v+1,\ w,\ l)$ and update.
- Start for Swim process:
  - a) Reset swim length counter to 0
  - b) Check for condition $m < N_s$
- Increment $m=m+1$
- If $J\ (u,\ v+1,\ w,\ l) < J_{last}$ set $J_{last=}J\ (u,\ v+1,\ w,\ l)$ and let

$$\theta^{u}(v+1,w,l) = \theta^{u}(v,w,l) + C(u)\frac{\Delta(u)}{\sqrt{\Delta^{T}(u)\Delta(u)}}$$

Using this $\theta^{u}\ (v+1,\ v,\ w)$ update $J\ (u,\ v+1,\ w,\ l)$.
Else, set $m=N_s$.
Increment $(u+1)$ if $u \neq S$

Step 5: Check for condition $v < N_c$, if yes, go to *Step 4.*

Step 6: In this step, the reproduction process starts:

for each bacteriam $u = 1, 2,....S$ calculate the health of bacterium as:

$$J^{u}_{health} = \sum_{v=1}^{N_c+1} J(u,v,w,l)$$

Sort the cost function values, J_{health}, in ascending order. The bacteria with the highest J_{health} values die and remaining bacteria with the best values split to progress the reproduction procedure.

Step 7: Check for condition $w < N_{re}$, if yes, go to *Step 3*.

Step 8: This step is the elimination–dispersal phase where with a defined probability bacteria are eliminated and dispersed to keep population constant).

Step 9: Check for condition $l < N_{ed}$, if yes, then go to *Step 2*; otherwise end.

The flowchart of BFO algorithm is given below (Fig. 2.13)

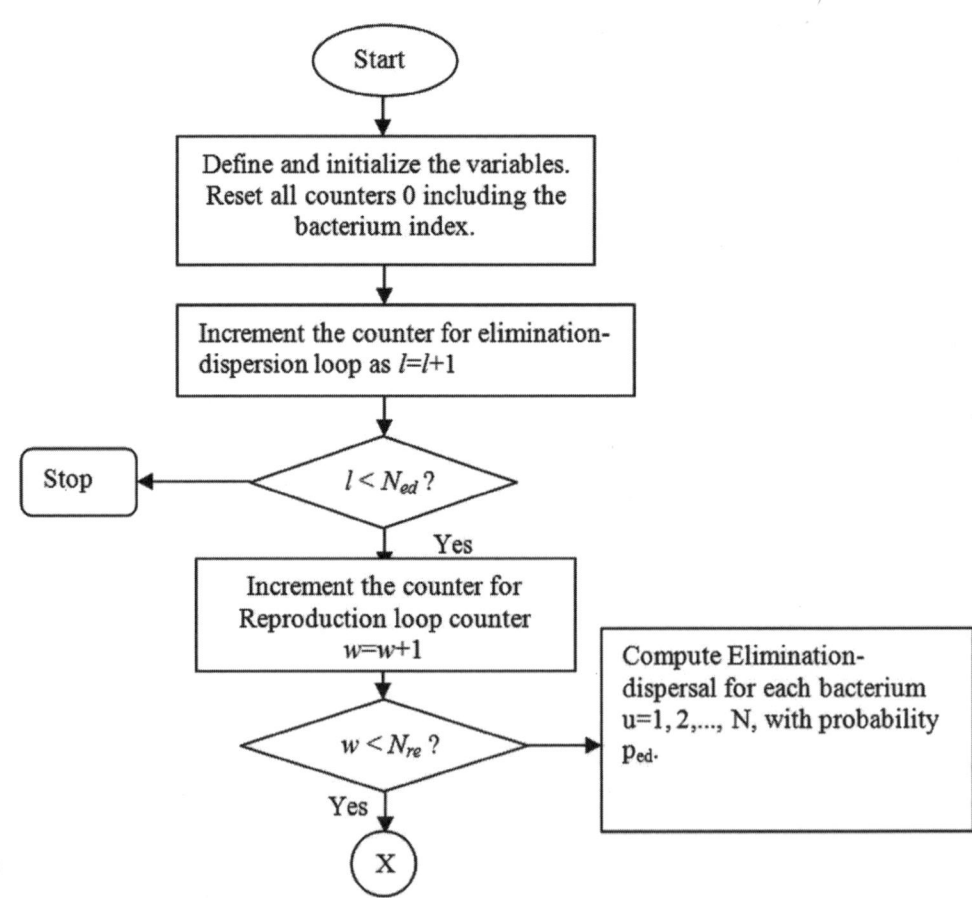

(i)

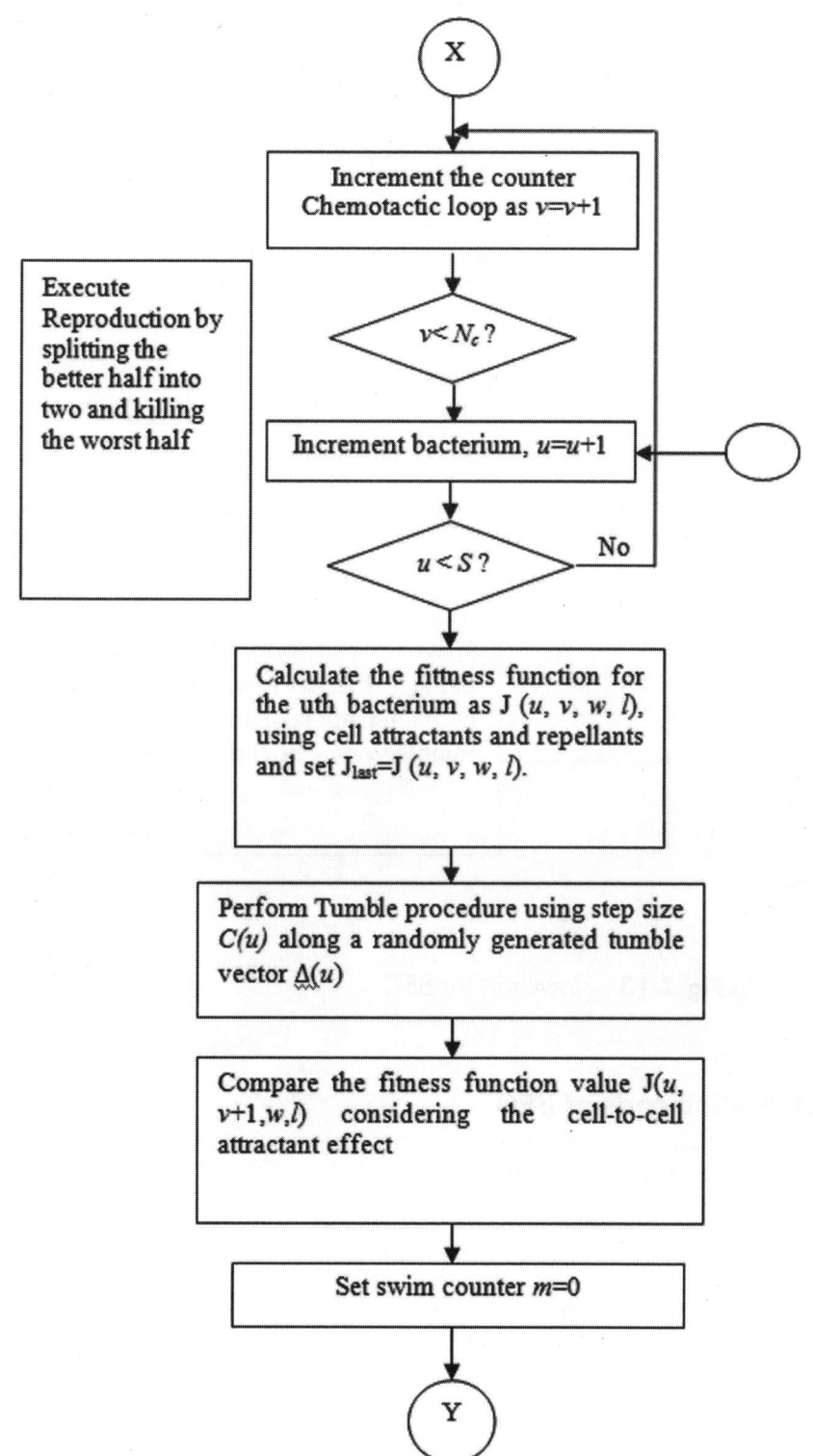

(ii)

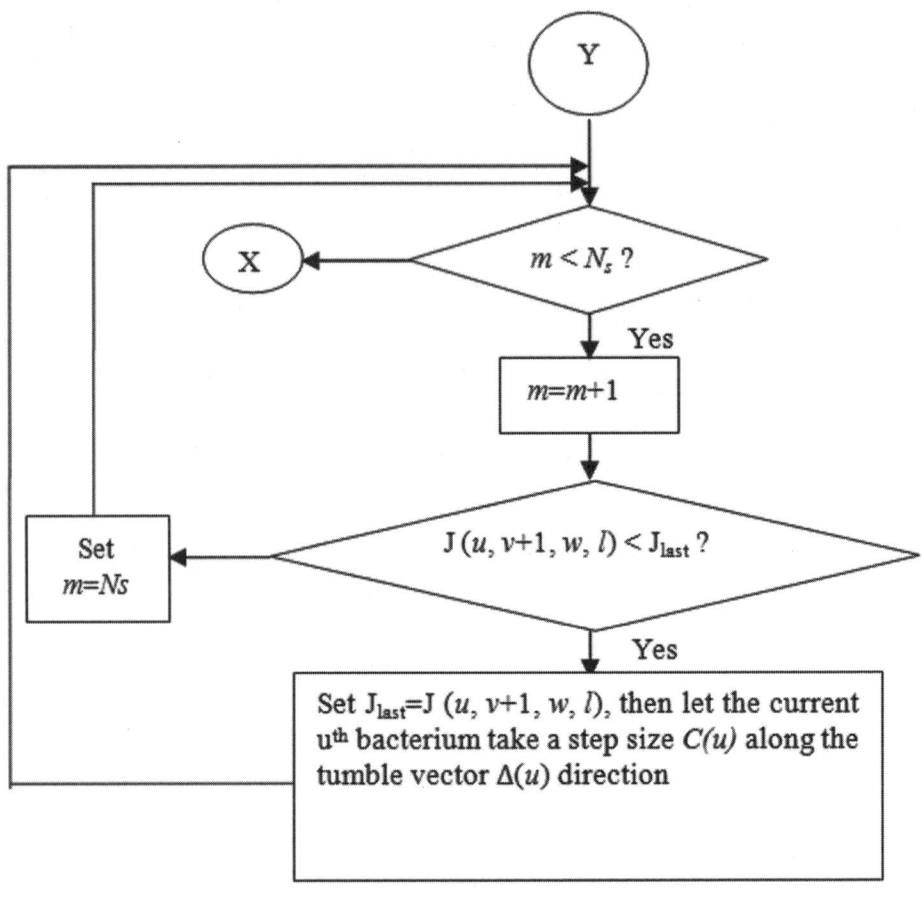

(iii)

Fig 2.13 Flowchart for BFO

2.4.4 Matlab code of BFO

```
clear all
clc
Nd=2;
Np=4;
Nc=66;
Ns=4;
Nre=4;
Ned=2;
Nsr=Np/2;
```

```
Prob_ed=0.25;
Rn_len(:,1)=0.05*ones(Np,1);

for m=1:Np  %Define Initial Positions
        P(1,:,1,1,1) = rand(Np,1)';
        P(2,:,1,1,1)= rand(Np,1)';
end
%-----Start Main Loop-----
%Loop for Elimination and dispersion
for ell=1:Ned
    %Reproduction
    for w=1:Nre

        for v=1:Nc
            for u=1:Np
            % display(P(:,u,v,w,ell));
                J(u,v,w,ell)=test_func(P(:,u,v,w,ell));
                % Tumble
                Jlast=J(u,v,w,ell);
                Delta(:,u)=(2*round(rand(Nd,1))-1).*rand(
Nd,1);
                P(:,u,v+1,w,ell)=P(:,u,v,w,ell)+Rn_
len(u,w)*Delta(:,u)/sqrt(Delta(:,u)'*Delta(:,u));

                J(u,v+1,w,ell)=tracklsq(P(:,u,v+1,w,ell));
                m=0;
                    while m<Ns
                        m=m+1;
                        if J(u,v+1,w,ell)<Jlast
                            Jlast=J(u,v+1,w,ell);
                            P(:,u,v+1,w,ell)=P(:,u,v+1,w,ell)
+Rn_len(u,w)*Delta(:,u)/sqrt(Delta(:,u)'*Delta(:,u)) ;
                            J(u,v+1,w,ell)=test_
func(P(:,u,v+1,w,ell));
                        else
                            m=Ns ;
                        end
                    end
                J(u,v,w,ell)=Jlast;
                sprintf('The value of iteration u %3.0f ,v =
%3.0f  , w= %3.0f, ell= %3.0f' , u, v, w ,ell );
```

```
            end
        end
%Reproduction
        Jhealth=sum(J(:,:,w,ell),2);
        [Jhealth,sortind]=sort(Jhealth);
        P(:,:,1,w+1,ell)=P(:,sortind,Nc+1,w,ell);
        Rn_len(:,w+1)=Rn_len(sortind,w);
            for u=1:Nsr
                P(:,u+Nsr,1,w+1,ell)=P(:,u,1,w+1,ell); % The
least fit do not reproduce, the most fit ones split into two
identical copies
                Rn_len(u+Nsr,w+1)=Rn_len(u,w+1);
            end
        end %  Go to next reproduction
        %Elimination and dispersal
        for m=1:Np
            if  Prob_ed>rand
                P(1,:,1,1,1) = rand(Np,1)';
                P(2,:,1,1,1)= rand(Np,1)';
            else
                P(:,m,1,1,ell+1)=P(:,m,1,Nre+1,ell);
            end
        end
    end % Gó to next elimination and dispersal
            reproduction = J(:,1:Nc,Nre,Ned);
            [vlastreproduction,O] = min(reproduction,[],2);  %
min cost function for each bacterial
            [Y,I] = min(vlastreproduction);
            pbest=P(:,I,O(I,:),w,ell);
```

2.5 Summary

This chapter introduces soft computing techniques and their algorithms for better understanding of the optimization problems described in the next chapters. The widely used algorithms such as neural networks, genetic algorithm, particle swarm optimization and bacteria foraging optimization have been described along with their terminologies, and algorithms. The algorithms are well explained through optimization of simple mathematical functions along with their Matlab code.

These algorithms emulate biological systems. Artificial neural networks try to replicate the pattern recognising behaviour of the brain. Genetic algorithm is based on the theory of evolution. Particle swarm optimization is inspired by societal behavioural pattern of a flock of birds and insect swarms, whereas bacterial forgaing optimization is developed from the replication of the food-ingesting behaviours of *escherichia coli* (*E. coli*) bacteria, present in intestines. These non-traditional optimization techniques are gaining popularity in every field of engineering problems. They are found to be potential optimization algorithms for complex optimization problems.

Classical optimization techniques have the following demerits:

- They are mostly time consuming.
- They fail to deal with complex problems involving several parameters.
- They are local search methods and involve derivative techniques.
- They are computationally unstable or less efficient.

These demerits are overcome by the above-mentioned soft-computing techniques. The effectiveness of the implementation of these techniques in the field of electromagnetics is discussed in the coming chapters.

References

Choudhury, B., S. Bisoyi, and R. M. Jha, "Emerging trends in soft computing for metamaterial design and optimization," *Computers, Materials & Continua,* vol. 31, no. 3, pp. 201–228, 2012.

Christodoulou, C., and M. Georgiopoulous, *Application of Neural Networks in Electromagnetics,* Nowrood, MA, Artech House, ISBN: 9780890068809, 512p., 2000.

Fieldsend, J. E., and S. Singh, "A multi-objective algorithm based upon particle swarm optimisation, an efficient data structure and turbulence," *Proceedings of UK Workshop on Computational Intelligence,* pp. 37–44, Sept. 2002.

Goldberg, D. E., *Genetic Algorithms in Search, Optimization and Machine Learning,* Addison-Wesley Publishing Company Inc., ISBN: 0-201-15767-5, 412p., 1989.

Haykins, S., *Neural Networks: A Comprehensive Foundation*, Prentice Hall International, NJ, ISBN: 9780139083853, 842p., 1999.

Holland, J. H., *Adaptation in natural and artificial systems*, University of Michigan Press, Ann Arbor, 1975.

Jang, J. S. R., C. T. Sun, and E. Mizutani, *Neuro-Fuzzy and Soft Computing*, Prentice Hall, NJ, ISBN: 9780132610667, 614p., 1997.

Jin, N., and Y. R. Samii, "Advances in particle swarm optimization for antenna designs: Real-number, binary, single-objective and multiobjective implementations," *IEEE Transactions on Antennas and Propagation*, vol. 55, no. 3, Mar 2007.

Kennedy, J., and R. Eberhart, "Particle swarm optimization," *Proc. of IEEE International Conference on Neural Networks,* pp. 1942–1948, 1995.

McCulloch, W. S., and W. Pitts, "A logical calculus of the ideas immanent in nervous activity," *Bulletin of Mathematical Biophysics,* vol. 5, pp. 115–133, 1943.

Mostaghim, S., and J. Teich, "Strategies for finding good local guides in multi-objective particle swarm optimization," *Proceedings of IEEE Swarm Intelligence Symposium,* pp. 26–33, Apr. 2003.

Passino, K. M., "Biomimicry of bacteria foraging for distributed optimization and control," *IEEE Control Systems Magazine*, vol. 22, pp. 52–67, 2002.

Passino, K. M., *Biomimicry for Optimization, Control and Automation*, Springer-Verlag, London, ISBN: 1-85233-804-0, 926p., 2005.

Patnaik, A., and R. K. Mishra, "ANN techniques in microwave engineering," *IEEE Microwave Magazine*, vol. 1, no. 1, pp. 55–60, March 2000.

Patnaik, A., D. Anagnostou, C. G. Christodoulou, J. C. Lyke: "Neurocomputational Analysis of a Multiband Reconfigurable Planar Antenna," *IEEE Trans. on Antennas and Propagation*, vol. 53, no. 11, pp. 3453–3458, Nov. 2005.

Samii Y. R., "Metamaterials in antenna applications: Classifications, designs and applications," *Proceedings of IEEE International Workshop on Antenna Technology, Small Antennas And Novel Metamaterials,* 2006, pp. 1–4, Mar. 2006.

Wasserman, D. P., *Advanced Methods in Neural Computing*. Van Nostrand Reinhold, New York, ISBN: 9780442004613, 255p., 1993.

Zadeh, L. A, "Fuzzy logic, neural networks and soft computing," One-page course announcement of CS 294-4, Spring 1993, *The University of California at Berkeley*, November 1992.

Soft Computing In Electromagnetics: A Review

3

Soft computing finds application in a wide range of problems in both engineering and non-engineering fields. Chapter 1 of this book discusses the potential applications of soft computing in fields ranging from engineering to finance, and architecture among others, etc. As the focus of this book is on design optimization of electromagnetic applications, it is necessary to understand the common optimization problems, and the advances and solutions to overcome them. Hence, a comprehensive review of soft computing techniques with a focus on electromagnetic applications is reported in this chapter.

3.1 Overview

An important aspect of electromagnetic applications is design and optimization towards actual hardware realization. In this chapter, an extensive literature survey of the soft computing techniques for electromagnetic applications has been carried out. It is observed that *artificial neural network* (ANN) and *genetic algorithm* (GA) has been employed extensively for diverse microwave engineering applications [Choudhury *et al.*, 2012]. In contrast, emerging soft computing techniques like *particle swarm optimization* (PSO) and *bacterial foraging optimization* (BFO) have not been explored comprehensively for these applications. Hence, soft computing techniques for various microwave engineering applications such as antenna engineering, frequency selective surfaces, radar absorber design applications, microwave devices, etc., are systematically reviewed in this chapter. This chapter also identifies the emerging trends and suitability of different soft computing techniques for various electromagnetic design and optimization problems.

3.2   Radar Absorbers

As the name suggests, electromagnetic absorbers are devices that absorb any incident radiation. In other words, the reflection off, and transmission through these devices is zero and the entire incident energy is absorbed by the materials present in the absorbers. The resonant properties of these absorbers are dependent on the constituent material, structure, and morphology. Conventionally, most absorbers are multi-layer in nature and consist of multiple dielectrics stacked one above the other. The thicknesses of these layers play an important role in the performance of the absorber. In addition, advances in the field of metamaterials have resulted in the use of metamaterials in absorber designs. A multi-band metamaterial absorber is given in Fig. 3.1 along with the absorption characteristic. It is seen that each peak in the absorption characteristic corresponds to a ring in the metamaterial structure [Shen *et al.*, 2011]. Therefore, it is clear that designing of radar absorbers is a complicated task and requires careful manipulation of material and structural properties. This necessitates the need for a robust, accurate, and quick optimization tool—a challenge that is met by soft computing techniques. In fact, numerous instances of the implementation of soft computing are available in literature, a few of which is discussed in this section.

The most popular amongst the soft computing techniques for radar absorber design has been the genetic algorithm (GA). Abundant literature is available on the implementation of GA for the optimization of metamaterial absorbers. Kollatou *et al.* [Kollatou *et al.*, 2011] implemented GA in order to design a metamaterial absorber for near unity absorption at 12.8 GHz. The optimization of cell parameters resulted in the development of a wide angle, enhanced bandwidth, and ultra-thin absorber.

Kern and Werner [Kern and Werner, 2003a] used GA to determine the structural parameters of a HZ–FSS (high impedance frequency selective surface). This HZ–FSS was used in the design

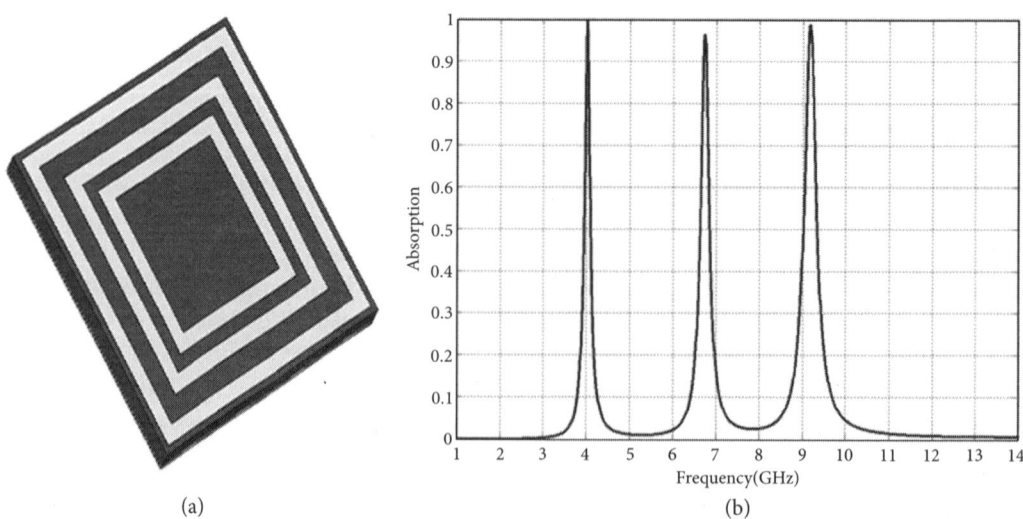

(a)   (b)

Fig 3.1   Schematic of a typical absorber, (a) multi-band metamaterial absorber unit cell, (b) absorption characteristics

of ultra-thin electromagnetic bandgap (EBG) absorbers for application in the microwave band. The GA aimed to optimize the fitness function (Eq. 3.1) by varying the dielectric constant, overall thickness and unit cell size of the HZ–FSS:

$$FF = \frac{1}{0.2|\varphi_{max}/180| + 0.8|\Gamma_{max}|} \tag{3.1}$$

where, φ_{max} and Γ_{max} are the maximum reflection coefficient magnitude and phase, respectively.

A similar strategy was presented by Liang et al. [Liang et al., 2005] for the design of ultra-thin EBG absorbers. In this case, the FSS was placed on top of a thin dielectric layer backed by a conducting polymer. A ground plane was then placed on the other side. GA was then implemented to obtain resonance at 4 GHz by changing the structural properties. Maximum absorption was observed at a conductivity of 40 S/m.

Wang et al. [Wang et al., 2008] proposed the implementation of GA in order to optimize metamaterial designs for enhanced bandwidth operation, i.e., for reflection of -12 dB over a bandwidth from 0.4 to 10 GHz. Specifically, the design of a two-layer absorber backed by a conductor was presented. The topmost layer was an optimized metamaterial layer that exhibited left-handedness. This layer was considered to be responsible for the enhanced bandwidth and was backed by a layer made up of a material with high absorptivity.

Another instance of optimization of FSS screens for absorber design is given in the work published by Wang and Werner [Wang and Werner, 2009]. GA was used for the analysis of a resistive FSS sheet for the development of a wide-band, double sided *electromagnetic band gap* (EBG) absorber. The formulation for the design was presented using an efficient spectral domain *periodic method of moments* (PMM). The resultant absorber showed wide-band and wide-angle response.

Bayraktar et al. [Bayraktar et al., 2010] discussed the design of a multi-layer metallo–dielectric absorber made of periodic, GA-optimized FSS screens sandwiched between dielectric layers. The FSS screen was designed to be electrically small. Reflection less than 45 dB was observed in this design.

A similar design was proposed by Jiang et al. [Jiang et al., 2010b] for absorber design. However, in this case, a dielectric layer was sandwiched between two metallic sheets—one consisting of patterns and the other being continuous. The geometry of the absorber was optimized using GA for absorptivity greater than 0.94 using the cost function:

$$\text{Cost} = \sum\nolimits_{freq} \cdot \sum\nolimits_{\theta_i} [|A_{i,TE} - A_{tar}| + |A_{i,TM} - A_{tar}|] \tag{3.2}$$

where, $A = 1 - |S_{11}|^2 - |S_{21}|^2$ and the desired absorption, i.e., $A_{tar} = 1$. $A_{i,TE}$ and $A_{i,TM}$ are the absorption for corresponding TE and TM polarization, respectively. The resultant design was seen to resonate in the mid-infrared band.

The popularity of soft computing for the optimization of absorbers with nano-structures is also seen in literature. Micheli et al. [Micheli et al., 2011] used genetic algorithm for designing a nano-structured multi-layer absorber. The same technique was also used to design shielding structures. The algorithm was flexible enough to enable selection of parameters such as frequency band, overall thickness, frequency range, etc. The user was given the choice of assigning priorities to thickness minimization and loss maximization. Similarly, a GA optimized doubly periodic array of stub-loaded H-shaped nano-particles was used to fabricate a conformal absorber [Jiang *et al.*, 2011]. This optimized absorber showed dual band characteristics in the infrared band with absorption greater than 90% at 3.3 μm and 3.9 μm.

Apart from metamaterials and FSS, another type of absorber made up of GA optimized electromagnetic smart screen (ESS) has also been proposed by Liu *et al.* [Liu *et al.*, 2011]. The algorithm maximized the absorption and the operational bandwidth. This ESS screen was simulated using finite element simulation and the reflection was measured using free space method. The relative dynamic bandwidth was observed to be as large as 40%.

3.3   Frequency Selective Surfaces

Metamaterial designs are static geometries. These geometries along with the spacing of unit cells determine the frequency response of the metamaterial. Due to this property, a new set of metamaterial unit cells must be added to the existing structure if the user wants to obtain a new response. Frequency selective surfaces (FSS) provide a viable alternative to this problem and allow for optional changes in the frequency response of the material itself. A typical FSS geometry is given in Fig. 3.2. Due to this desirable property, FSS finds application in conformal antennas, radomes, etc. However, just like a metamaterial, an FSS is also multi-layered in nature and soft computing is known to be used for design optimization.

Mumcu *et al.* [Mumcu *et al.*, 2007] used genetic algorithm in order to optimize the transmission response of an FSS structure. On comparison with the ideal (expected) response, the resonance of the FSS was found to match perfectly. This intensive algorithm used 24 distributed computers in order to design complicated composite material textures and display unique dispersion and field characteristics. The total run time of the algorithm for obtaining the design was found to be 11 hours.

Fig 3.2 Typical frequency selective surface

The genetic algorithm was also used by Gingrich and Werner [Gingrich and Werner, 2005b] in order to design a planar, thin, low-loss, zero index metamaterial (ZIM) using metamaterial. The algorithm was given design constraints of centre frequency, total loss required, and refractive index, and it optimized the FSS and substrate/superstrate simultaneously.

Kern and Werner [Kern and Werner, 2003b] demonstrated the optimization of the surface impedance of a high impedance FSS (HZ–FSS) in order to obtain the desired value of permittivity—both real and complex. The soft computing technique that was used was the genetic algorithm.

Kern *et al.* [Kern *et al.*, 2003c] also designed a multi-frequency artificial magnetic conductor (AMC) by placing an FSS over a thin dielectric layer followed by a perfect electric conductor. The entire assembly was optimized using genetic algorithm in order to show narrow-band resonance at multiple resonance bands, viz. 8 and 23 GHz (dual band resonator), and 3.5, 11 and 18 GHz (triple band resonator). Similarly, Ge and Esselle [Ge and Esselle, 2007] combined micro-genetic algorithm with finite difference time-domain method (FDTD) in order to design single and multi-band AMC surfaces based on FSS screens.

Bossard *et al.* [Bossard *et al.*, 2008] used the genetic algorithm to design an FSS with nematic liquid crystal superstrate. The target was to achieve the characteristics of a broadband, tunable, mid-IR filter, which finds application in the design of IR/optical switches. The cost function used for optimizing the superstate was:

$$C = \sum_{Pass} R^2 + \frac{3}{2} \sum_{Stop} T^2 \tag{3.3}$$

where, R and T are reflectance and transmittance, at each specified pass-band and stop-band frequency, respectively.

3.4   Antenna Design and Optimization

Another interesting application of soft computing techniques is in the area of antenna design and optimization. Typically, the soft computing technique is used to achieve one or more of the following goals: antenna miniaturization, antenna pattern synthesis, and performance enhancement.

3.4.1 Antenna miniaturization

The wireless communication community always demands high performance and miniaturized antennas. Planar Inverted F antenna (PIFA) is a popular antenna used in mobile communication. The size of this kind of antennas can be reduced through proper optimization by embedding a metamaterial structure into it. A typical metamaterial PIFA system is shown in Fig. 3.3.

Xin-yuan *et al.* [Xin-yuan *et al.*, 2011] proposed a novel design of metamaterial loaded wideband PIFA system. An array of Jerusalem cross-shaped PEC structure used in frequency selective surface (FSS) was designed to act as a metamaterial structure. This metamaterial structure was loaded on the ground plane of the PIFA antenna. The metamaterial loaded PIFA antenna enhanced the bandwidth by 49.4% from 2.56 GHz to 4.24 GHz while the size of the antenna also reduced by 91%.

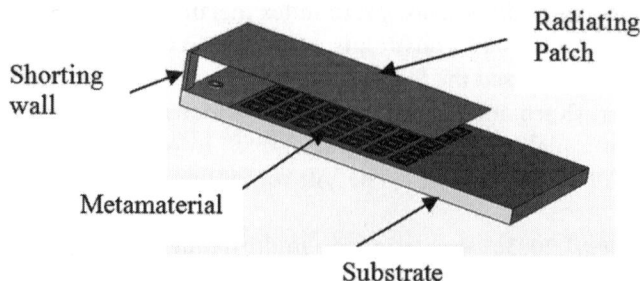

Fig 3.3 Schematic of a metamaterial loaded planar inverted F antenna (PIFA)

Azadegan and Sarabandi [Azadegan and Sarabandi, 2003] used genetic algorithm in conjunction with full wave forward model towards miniaturization of slot antennas. Equivalent circuit model was used for antenna and feed structure design and optimization. The corresponding simulation and measurement results of a prototype slot antenna were provided. The gain was found to be 3dBi. Further, for a small ground plane, the antenna resonated at 300 MHz.

Viani *et al.* [Viani *et al.*, 2012] used particle swarm optimization for design optimization of a multiband compact fractal antenna that operates in the UHF and GPS-L1 band. A prototype optimized antenna was fabricated. The simulated and measured results were reported, which showed that the two are in excellent agreement with each other.

Further miniaturization of a Yagi–Uda array was carried out by Bayraktar *et al.* [Bayraktar *et al.*, 2006] using particle swarm optimization. The optimized results show that the length and width of the array elements can be reduced from 70% to 44 % through proper optimization techniques.

Werner *et al.* [Werner *et al.*, 2005] proposed design optimization of a multiband monopole antenna. Genetic algorithm was used for size reduction of upto 30% without disturbing the multiple resonance. The algorithm was proved through simulation and design of a miniaturized dual band monopole antenna and a tri-band whip antenna.

Assimonis *et al.* [Assimonis *et al.*, 2012] used a computation FEM-based eigenvalue analysis in order to design a uni-planar compact cross-like EBG structure. The design found potential application in MIMO antennas with reduced coupling when GA was used to maximize the bandgap zone.

PSO based quadrature reflection phase structure topology determination was presented by Jin and Samii [Jin and Samii, 2005]. The PSO optimizer was integrated with FDTD in order to obtain miniaturization at 5.7 GHz and 8.8 GHz.

3.4.2 Antenna pattern synthesis

Active antenna arrays are gaining momentum in the field of antenna engineering because of their capabilities including re-configurability, beam steering, and adaptive nulling. Soft computing techniques may be used to synthesize the radiation pattern of an antenna array by specifying the maximum allowable side lobe level for a range of angles. For example, consider that a design requires the sidelobe levels (SLL) of a non-uniform antenna array in the angular ranges from 0 to 80° and 100° to 180° to be less than −18 dB. The distances between the elements are varied by particle swarm optimization in order to achieve this. The variation of the fitness function with respect to iteration number and the radiation pattern of the final array is given in Figs. 3.4 and 3.5.

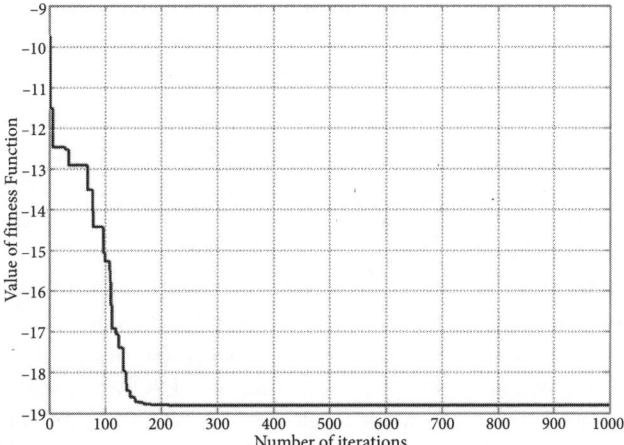

Fig 3.4   Variation of fitness function versus number of iterations for reduction in sidelobe level of antenna array

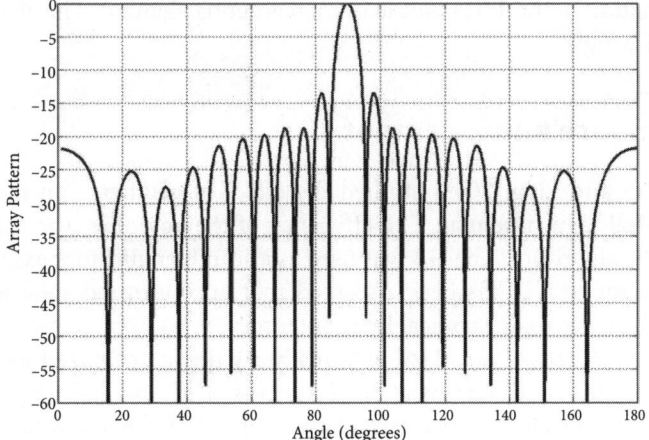

Fig 3.5   Radiation pattern after optimization with desired side-lobe level for the desired angles

Jafargholi and Kamyab [Jafargholi and Kamyab, 2010] proposed an ultra-wide band (UWB) array by placing array elements on a spiral curve. The aim was to improve the array performance such as the array bandwidth, radiation pattern, directivity, and front-to-back ratio, using soft computing. Specifically, the radiation pattern was shaped by the genetic algorithm in order to be suitable for UWB operation. It was observed that a metamaterial covering improved directivity by 5–7 dB.

Kim and Yeo [Kim and Yeo, 2007] designed and fabricated a dual band passive RFID tag antenna using an artificial magnetic conductor (AMC) ground plane, whose lumped element in the equivalent circuit was optimized using a hybrid genetic algorithm. The designed antenna exhibited dual band operation at European and Korean UHF frequency bands.

Scarborough et al. [Scarborough et al., 2011] designed a low-cost, dual polarization, horn antenna, which included a metamaterial lining. GA was used to arrive at low sidelobes and low cross polarization. It was seen that the radiation patterns were almost independent of polarization.

Yaman and Yilmaz [Yaman and Yilmaz, 2010] explored the effect of crossover and mutation rate on the performance of real coded genetic algorithm. The analysis was performed via a circular antenna array pattern synthesis problem. It has been observed that low mutation rate shows good performance whereas the restricted cross-over scheme provides better result than the random crossover scheme.

Acharya et al. [Acharya et al., 2011] used PSO for suppression of sidelobe level of faulty antenna arrays while maintaining the nulls at the desired level. For certain applications, such as space platforms and deployment in battle fields, replacement of a faulty array is not feasible. The proposed technique will be feasible in such situations. PSO was used here to optimize the amplitude and phase excitation of the working elements in order to extract the original pattern. The algorithm was tested using a 20 element linear Chebyshev array with -30 db side lobe level and null at 20°.

Kumar and Singh [Kumar and Singh, 2013] used real coded genetic algorithm for phase only pattern synthesis of active phased array where the beam pattern can be broadened by changing the excitation in the T/R module of each antenna element. The algorithm was tested for a planar dipole array.

3.4.3 Performance enhancement

Lim and Ling [Lim and Ling, 2009] studied the effect of geometry on the performance of spherical and conical helix antennas. The efficiency of the antennas was enhanced to 87.5 % using micro-genetic algorithm. The GA was used as a search engine to maximize the efficiency with an optimal geometry. It has been observed that an optimized spherical helix antenna shows better performance than a conical helix antenna.

Similarly, a 2 × 2 circularly polarized antenna array being optimized using micro-genetic algorithm was presented by Kossiavas et al. [Kossiavas et al., 2011]. The performance of the antenna was improved using a metamaterial based partially reflective surface (PRS). The said design was subsequently fabricated and measured, which operated in the frequency band between 9–9.8 GHz.

A novel stacked-patch antenna was designed by Werner et al. [Werner et al., 2009], which was made up of matched magneto-dielectric materials as substrates. Genetic algorithm along with periodic moment–method (PMM) code was used to select the optimal material thicknesses, dielectric constant, FSS screen geometry and the unit cell size to reduce the total thickness of the antenna. The targeted centre frequency of the antenna was 1.2 GHz and bandwidth 20%, with VSWR less than 2:1.

Kossiavas and Dubard [Kossiavas and Dubard, 2007] proposed a Fabry–Perot cavity antenna that behaved like an AMC. The Fabry–Perot cavity consisted of a perfect electric conductor (PEC), a partially reflecting surface (PRS) and a patch antenna. The design requirement was to simultaneously improve the antenna gain and decrease the total thickness for operation in the UMTS–WLAN frequency band (1.95 GHz; 2.5 GHz) using micro-genetic algorithm

Chen *et al.* [Chen *et al.*, 2008] used the genetic algorithm in order to design an artificial magnetic metamaterial. The structure was generated via a technique called filling element methodology. GA optimized the structure in order obtain negative permeability using the following fitness function:

$$\text{fitness} = \min \left(\left| \left(\mu'_{eff} + R_a \right) \right| + \left| \left(\mu''_{eff} + R_b \right) \right| \right) \Big|_{4GHz < f < 10GHz} \qquad (3.4)$$

where, μ'_{eff} and μ''_{eff} represents the real and imaginary part of effective permeability, μ_{eff}. For permeability of -1, R_a =1 and R_b=0.

Tonn and Bansal [Tonn and Bansal, 2009] used GA to enhance the performance of a HF band (3–30 MHz) linear insulated, metamaterial coated antenna. The performance attributes that were optimized, were bandwidth, gain, and gain-bandwidth product. The presence of the metamaterial coating was explained using a dispersion model and it was shown that the resultant structure behaved like a capacitive loaded antenna.

3.5   Metamaterial Structures

Metamaterials are artificially engineered materials that show unusual properties such as negative permittivity, permeability and negative refractive index, etc. These structures typically consist of metallic patterns etched onto substrates and are realized through split ring resonators, artificial magnetic conductors, etc. Metamaterials can also be realized using three-dimensional structures. One such structure proposed by Moser *et al.* [Moser *et al.*, 2009] is the meta-foil (Fig. 3.6). Metamaterials are frequency specific in nature and the resonant frequency depends

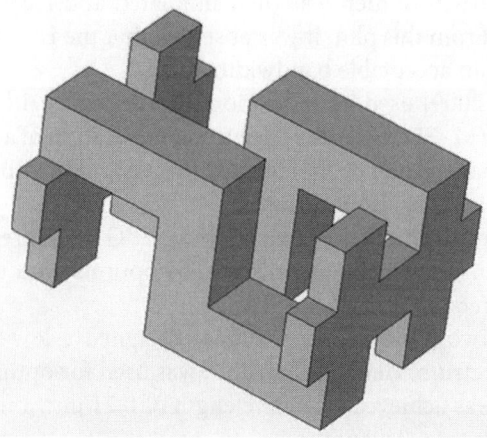

Fig 3.6 A typical 3D structure of a metamaterial unit cell used as meta-foil

on the dimensions of the metallic structure, thickness and material of the substrate. As a result, for proper operation at a particular frequency, it becomes imperative to find the optimum dimensions of the metamaterial. Therefore, just like FSS discussed previously, soft computing has also been proven to be extremely efficient for design of metamaterial structures.

Kern et al. [Kern et al., 2005] used genetic algorithm for the development of meta-ferrites having desirable magnetic properties at frequencies above 1 GHz. The algorithm was used to optimize HZ–FSS and achieve the required effective permeability. Three FSS structures were analyzed and the final design was optimized using the fitness function:

$$FF = -\sum_{n=1}^{N_f} \left(\mu_r'(f_n) - 4\right)^2 + \left(\mu_r''(f_n) - 3\right)^2 \tag{3.5}$$

where, N_f is the number of frequencies considered for optimization.

Gingrich and Werner [Gingrich and Werner, 2005a] presented a novel method for synthesis of planar, thin, low loss, light weight FSS using GA. For low/ zero refractive index metamaterial (LIM/ZIM), the operational goals included bandwidth, frequency, surface impedance, effective refractive index, and transmission. Over 8–10 GHz, The real part of the inverted effective index of refraction of the optimized structure was very low (0.005).

Kwon et al. [Kwon et al., 2007c] optimized the parameters of the low/ zero refractive index metamaterial (LIM/ZIM) using GA. A near zero permittivity was achieved through a checkerboard pattern generated with the use of dielectric and Drude type materials. The checkerboard patterns were stacked one over the other in order to realize the metamaterial. The genetic algorithm optimized this stacked metamaterial structure for a desired value of permittivity and permeability.

Zhao et al. [Zhao et al., 2011] developed a 20 GHz, negative index metamaterial with desired electromagnetic characteristics using metamaterial microstructure design methodology. The design procedure involved the use of an FDTD solver and genetic algorithm in conjunction with each other.

Werner et al. [Werner et al., 2005] used GA to optimize a dual-band GPS/cellular EBG AMC FSS ground plane. The structure then was then simulated and fabricated, and the reflection phase plot was obtained. From this plot, it was observed that the EBG resonated at the targeted resonant frequency, with an acceptable bandwidth.

Kim et al. [Kim et al., 2009] used hybridization of genetic algorithm along with multilayer finite–difference method (M–FDM) for the design and fabrication of an EBG structure. The GA algorithm generated a binary sequence that formed the basis of the EBG design. Verification of this method through Matlab was also illustrated.

Further, bandwidth improvement of planar periodic EBG structure was carried out by Deias et al. [Deias et al., 2009] using genetic algorithm. The optimization was used to find the best structure for a resonant frequency of 14.25 GHz.

Kwon and Werner [Kwon and Werner, 2007a] designed a low-index metamaterial that operates in the visible spectrum. Genetic algorithm was used for optimization of the structure. A transmittance of 68% was achieved at a wavelength of 0.71 μm, and a transmittance of 88% was achieved at a wavelength of 0.69 μm.

Bayraktar et al. [Bayraktar et al., 2007, 2009b] proposed a technique to synthesize an artificial magnetic conductor (AMC) metamaterial structure made of a dielectric sandwiched between

FSS screens, separated by a dielectric material. The top and bottom surfaces were optimized by a genetic algorithm. The structure showed dual band behaviour from 2.14 GHz to 2.23 GHz, and from 5.17 GHz to 5.32 GHz, with gains of 9 dB and 7 dB, respectively.

Ge and Esselle [Ge and Esselle, 2007] proposed utilising grounded FSS structures towards realization of single and dual band AMC surfaces. The FDTD design was optimized by using GA. The parameters such as thickness of substrate, permittivity, metallic pattern of the single band AMC structure were optimized to operate at 4.5 GHz–5.5GHz. Similar parameters were optimized for the dual band AMC to operate at 5.0 GHz–6.0 GHz and 11.5 GHz–12.5 GHz.

Chiral metamaterials were also designed and optimized using genetic algorithm [Kwon et al., 2008]. The unit cell of the double-periodic metamaterial design was optimized for maximum circular dichroism (CD) and optical cavity. Circular dichroism value of 56% was obtained at a wavelength of 1.087 μm.

A flexible low loss negative index metamaterial was designed by Bossard et al. [Bossard et al., 2009] consisting of stacked metal (gold) and dielectric screens with a periodic array of holes. The structure was optimized by genetic algorithm for the given refractive index (–1) with minimum absorption and impedance values in the mid-IR spectrum (85 THz–90 THz).

Kahlout and Kiziltas [Kahlout and Kiziltas, 2009] proposed a unit cell development technique to design and fabricate a heterogeneous periodic material through material simulation model and genetic algorithm. The EM material properties were achieved through composition of the developed heterogeneous material unit cell. Further, the design was optimized using a GA based frame-work.

A low loss planar multilayer metamaterial (NIM–ZIM–UIM) was designed and optimized by Jiang et al. [Jiang et al., 2009b]. The designed material consists of a fishnet metal screen sandwiched between photonic band-gap substrates. The parameters such as effective refractive index and free space impedance matching at desired frequency were optimized using genetic algorithm. Further, improvement on similar applications was reported by Jiang et al. [Jiang et al., 2010a] using genetic algorithm combined with a generalized inversion method. The designed and optimized structure showed superior absorption and impedance match.

Automatic retrieval of material properties such as effective permittivity and permeability of a metamaterial sample for a given frequency range through genetic algorithm in conjunction with standard transmission/reflection (TR) techniques was carried out by Weikai et al., 2010.

A genetic algorithm based CAD package was designed by Pradeep et al. [Pradeep et al., 2011] for design and synthesis of SRR, which exhibited negative magnetic permeability. The resonant frequency was in the range of 1–8 GHz.

Genetic algorithm was used by Radovanovic et al. [Radovanovic et al., 2011] to develop carrier transport model using quantum cascade lasers. The other applications of quantum cascade lasers such as amplification via inter sub-band transitions are also reported.

Similarly, it is possible to obtain spontaneous emission patterns of both the electric and magnetic dipoles. Ni et al. [Ni et al., 2011] used the dyadic Green's function to project these patterns on various metallic and even hyperbolic metamaterial surfaces through a pattern search and genetic algorithm.

Kwon et al. [Kwon et al., 2007] implemented genetic algorithm for design optimization of a 2-D metamaterial in the IR–visible spectrum. A comparison study of GA with PSO was carried out for convergence time. For this design example, PSO converged more quickly than the GA.

Tavallaee and Samii [Tavallaee and Samii 2007] developed an optimization algorithm that used multi-conductor transmission line in conjunction with particle swarm optimization (MTL–PSO) for electromagnetic bandgap power distribution network (EBG–PDN) design. The developed algorithm was used for optimization of a mushroom-like EBG structure. Using simple circuit models, the optimization time was reduced and accuracy comparable to full-wave simulators was maintained. The MTL–PSO integrated a 1-D MTL–PSO algorithm that reduced the EBG–PDN optimization time greatly with accuracy comparable to FE-based optimizer.

Another novel soft computing technique based on foraging behaviour of bacteria (BFO) was also used by several researchers for antenna engineering, and electronic control applications. But BFO is yet to be explored for metamaterial applications. Choudhury *et al.* [Choudhury *et al.*, 2013] implemented BFO for design and optimization of metamaterial structures. BFO is implemented to optimize the structural parameters of a SSRR at a desired frequency of operation. The BFO optimizer acts here as a CAD package, which yields the structural parameters like length (*a*), width (*w*), and spacing (*d*) at a desired resonant frequency. The cost function used for this optimization is

$$ f_{err} = \frac{|f_d - f_c|}{f_d} \tag{3.6} $$

where f_d is the desired frequency and f_c is the frequency calculated by equivalent circuit analysis. The BFO-optimized value, along with the equivalent circuit analysis method, is compared and the error is shown, which is within the tolerable limits (Table 3.1).

Table 3.1 Optimized structural parameter for desired frequency

Desired frequency (GHz) f_d	f_{err}	Design optimization output		
		Length (mm) *a*	Width (mm) *w*	Gap (mm) *d*
0.649	0.000074	15.4	0.7	0.5
0.879	0.000001	9.0	0.4	0.1
1.430	0.000007	6.4	0.6	0.1
3.362	0.000140	3.0	0.1	0.1

The capability of artificial neural network (ANN) for design and analysis of metamaterial structures was reported by Subramanian *et al.* [Subramanian *et al.*, 2012]. ANN was used for design optimization of dual log-spiral resonator (DLSR), which was an optimal design in the X band. The equivalent circuit model was used for data generation and network training. Although ANN takes significant computational time, it becomes effective once this training period is over.

Another metamaterial structure, (SSRR) was optimized using genetic algorithm (GA), artificial neural network (ANN) and hybrid GA–ANN by Vidyalakshmi and Raghavan [Vidyalakshmi and Raghavan , 2010]. In this work, the structural parameters are achieved by optimizing the resonant frequency. A comparative study of these three techniques were implemented and it was observed that GA provides a better search space and accuracy, the neural network provides quick solutions and hence the combination i.e., hybrid GA–ANN produced an effective structure.

3.6   Invisibility Cloaks

Invisibility cloaks are designed with the aim of hiding objects from view. In a practical sense, the technology of invisibility cloaks has been extensively used for radar avoidance while the same for hiding the wearer from other people (operation in the domain of visible light) still remains a hypothetical device. The reason for this restriction on useable frequency range lies in the nature of construction of cloaks; as of now technology has not advanced to a stage where visible light cloaking is feasible.

The concealment of an object is achieved when the incident radiation is not allowed to reflect or scatter. For best performance, the incident light must be forced to bend around the object and then resume its original path as if the object is not there in the first place.

This behaviour is achieved using the principle of transformation optics [Ivsic et al., 2011]. Maxwell's equations are subjected to a conformal co-ordinate transformation in order to obtain a spatially distributed set of constitutive parameters. These parameters define the cloak. For perfect cloaking, the material of the cloak should be anisotropic, spatially invariant and inhomogeneous in nature; the permittivity and permeability tensors of the cloak material are designed in this regard. In order to attain the aforementioned properties, the cloaks are often multi-layer in nature. The optimization of these multi-layer designs is possible through soft computing

Ivsic et al. [Ivsic et al., 2010] used particle swarm optimization in order to design a metamaterial based invisibility cloak. The algorithm optimized the parameters of the metamaterial layer in order to achieve minimum possible scattering width. It was reported that practical cloaks are often made of 10 layers. However, comparable performance may be obtained using three layers also.

Later Ivsic et al. [Ivsic et al., 2011] conducted the analysis of the bi-static RCS of cylindrical and spherical structures. Back-scattering was reduced by using particle swarm optimization and comprehensive learning PSO (CLPSO) in order to find the structural parameters of the cloak. A comparison between the performance of the two algorithms were provided and it is seen that the CLPSO required lesser number of particles (40 particles) as opposed to normal PSO (300 particles).

The genetic algorithm was used again by Jiang et al. [Jiang et al., 2009a] in order to optimize two multilayer ZIM cloaks based on FSS. Fabrication constraints were also included in the design. The aim of the GA was to optimize refractive index as well as impedance. Consequently, the following fitness function was used:

$$C = \left| n_{eff} - n_{\text{target}} \right|^2 + \left| z_{eff} - z_{\text{target}} \right|^2 \qquad (3.7)$$

where, n_{target} denotes desired refractive index and z_{target} is the free-space impedance. After optimization, the transmission loss at 11 GHz was found to be -0.08 dB.

Invisibility cloaks can also be designed using isotropic layered media. Feng et al. [Feng et al. , 2011] used GA to optimize the layer of one such cloak structure. The concept was used by Yu et al. [Yu et al., 2011] to develop a cylindrical cloak with minimum layers of non-negative isotropic materials. Using GA, a 20 dB reduction in scattering width in all dimensions was observed.

The genetic algorithm has also been used in the design of conformal cloaks. Specifically, Qui *et al.* [Qui *et al.*, 2008] used this algorithm to design and optimize a cylindrical cloak. The work also studied the EM and material properties of arbitrary cloaks and the effect of concentrators on cylindrical cloaks. Following this, improved cloaking models were provided. It was reported that the zeroth-order scattering could be eliminated completely for a certain wavelength by placing an additional layer at the inner surface of the simplified cloak. Further, Xu *et al.* [Xu *et al.*, 2011] used genetic algorithm for the RCS reduction of a broadband cylindrical cloak. The analysis for the same was carried out in free space and the optimization algorithm attempted to arrive at the best values for the finite constitutive parameters of the cloak for a resonant frequency of 2.12 GHz. A more advanced conformal cloak design technique was reported by Oraizi and Abdolali [Oraizi and Abdolali, 2008], in which a method of least squares (MLS) was developed by combining genetic algorithm and conjugate gradient method (CG). The MLS was used to control the RCS in an ultra-wide bandwidth. This was achieved by studying scattering of electromagnetic waves over multi-layered cylindrical structures.

The same technique, i.e., MLS was used again by Oraizi *et al.* [Oraizi *et al.*, 2010] in order to minimize reflected power in radar absorbing applications by optimizing double-zero metamaterials (DZR) coatings.

3.7   Microwave Devices

Soft computing can also be used to optimize microwave devices like transmission lines, stubs, etc. Gunnel [Gunnel, 2007] developed a continuous parameter genetic algorithm (CPGA) based mathematical model for composite right/left-handed (CRHL) non reciprocal, non-symmetric transmission line radial stub. A comparison between the algorithm used and the standard binary genetic algorithm (BGA) was also provided. It was concluded that CPGA was faster and required lesser memory than binary genetic algorithm (BGA). The same researcher, Gunnel [Gunnel, 2011] also used GA for the design of a dual-frequency, transmission-line impedance matching section. The synthesis was performed for both uniform and non-uniform matching sections.

Analysis and design of various transmission lines were also conducted using artificial neural networks. Patnaik *et al.* [Patnaik *et al.*, 2009] used ANN for design and analysis of microstrip transmission line and slot line transmission line, which has been discussed in detail in Chapter 6 of this book.

3.8   Summary

To summarize, this chapter provides the trends of soft computing field in electromagnetics including metamaterial absorber, FSS, antennas, cloaking, microwave devices, etc. This extensive literature survey reveals that artificial neural network and genetic algorithm is well known and has been widely used by the electromagnetics community. In contrast, the other well established soft computing techniques like PSO and BFO are yet to be explored for various applications of microwave engineering. The chapter also reports implementation of some of these emerging soft computing techniques (e.g. PSO, BFO) for various applications along with the comments on suitability of soft computing techniques for complicated EM design applications.

It has been also observed that hybridized soft computing techniques such as GA-ANN, MOPSO etc. will reduce the computation time and resources significantly. Further, it is there is a need to develop computational engines, which can integrate the optimization tool with certified EM solvers to achieve high accuracy designs.

References

Acharya, O. P., A. Patnaik, and S. N. Sinha, "Null steering in failed antenna arrays," *Applied Computational Intelligence and Soft Computing*, vol. 2011, pp. 1–9, 2011.

Assimonis, S. D., T. V. Yioultsis and C. S. Antonopoulos, "Computational investigation and design of planar EBG structures for coupling reduction in antenna applications," *IEEE Transactions on Magnetics*, vol. 48, no. 2, pp. 771–774, Feb. 2012.

Azadegan, R., and K. Sarabandi, "A novel approach for miniaturization of slot antennas," *IEEE transactions on Antennas and Propagation*, vol. 51, no. 3, pp. 421–429, Mar. 2003.

Bayraktar, Z., J. Bossard and D. H. Werner, "AMC metamaterials for low-profile antennas mounted on or embedded in composite platforms," *Proceedings of IEEE Antennas and Propagation Society International Symposium*, pp. 1305–1308, Jun. 2007.

Bayraktar, Z., M. Gregory and D. H. Werner, "Composite planar double-sided AMC surfaces for MIMO applications," *Proceedings of IEEE Antennas and Propagation Society International Symposium*, pp. 1–4, Jun. 2009*b*.

Bayraktar, Z., P. L. Werner, D. H. Werner, R. Zadegan, and K. Sarabandi, "The design of miniature three-element stochastic Yagi-Uda arrays using particle swarm optimization," *IEEE Antennas and Wireless Propagation Letters*, vol. 5, pp. 22–26, 2006.

Bayraktar, Z., X. Wang and D. H. Werner, "Thin composite matched impedance magneto-dielectric metamaterial absorbers," *Proceedings of IEEE Antennas and Propagation Society International Symposium*, pp. 1–4, Jul. 2010.

Bossard, J. A., S. Yun, D. H. Werner and T. S. Mayer, "Synthesizing low loss negative index metamaterial stacks for the mid-infrared using genetic algorithms," *Optics Express*, vol. 17, no. 17, pp. 14771–14779, Aug. 2009.

Bossard, J. A., X. Liang, L. Li, S. Yun, D. H. Werner, B. Weiner, T. S. Mayer, P. F. Cristman, A. Diaz and I. C. Khoo, "Tunable frequency selective surfaces and negative-zero-positive index metamaterials based on liquid crystals," *IEEE Transactions on Antennas and Propagation*, vol. 56, no. 5, pp. 1308–1319, May. 2008.

Chen, P. Y., C. H. Chen, H. Wang, J. H. Tsai and W. X. Ni, "Synthesis design of artificial magnetic metamaterials using a genetic algorithm," *Optics Express*, vol. 16, no.17, pp. 12806–12818, Aug. 2008.

Choudhury, B., S. Bisoyi and R.M. Jha, "Bacteria foraging algorithm for metamaterial design and optimization," 2013 *IEEE Applied Electromagnetics Conference* (AEMC), Bhubaneswar, India, Paper No.: CAD-2–1884, 2 p., December 18th –20nd, 2013.

Choudhury, B., S. Bisoyi and R.M. Jha, "Emerging trends in soft computing for metamaterial design and optimization," *Computers, Materials & Continua*, vol. 31, no. 3, pp. 201–228, 2012.

Deias, L., G. Mazzarella and N. Sirena, "Synthesis of EBG surfaces using evolutionary optimization algorithms," *Proceedings of European Conference on Antennas and Propagation,* pp. 99–102, Mar. 2009.

Feng, Y. J., X. F. Xu and Z. Z. Yu, "Practical realization of transformation-optics designed invisibility cloak through layered structures," pp. 3456–3460, 2011.

Ge, Y., and K. P. Esselle, "GA/FDTD technique for the design and optimization of periodic metamaterials," *IET Microwave Antennas propagation, pp. 158-164,* 2007.

Gingrich, M. A. and D. H. Werner, "Synthesis of low / zero index of refraction metamaterials from frequency selective surfaces using genetic algorithms," *Electronics Letters,* vol. 41, no. 23, Nov. 2005a.

Gingrich, M. A. and D. H. Werner, "Synthesis of zero index of refraction metamaterials via frequency-selective surfaces using genetic algorithms," *Proceedings of IEEE Antennas and Propagation Society International Symposium,* vol. 1A, pp. 713–716, Jul. 2005b.

Goudos, S. K. and J. N. Sahalos, "Microwave absorber optimal design using multi-objective particle swarm optimization," *Microwave and Optical technology letters,* vol. 48, no. 8, pp. 1553–1558, Aug. 2006.

Gunel, T., "Synthesis of a novel composite right/left-handed nonreciprocal and nonsymmetric transmission line radial stub," *Applied Electronics, AE* 2009, pp. 119–122, Sep. 2009.

Gunel, T., "Dual-frequency transmission line impedance matching sections," *International Conference on Applied Electronics* 2011, pp. 1–4, Sep. 2011.

Ivsic, B., T. Komljenovic, and Z. Sipus, "Time and frequency domain analysis of uniaxial multilayer cylinders used for invisible cloak realization," *Proceedings of conference on ICECom* 2010, pp. 1–5, Sep. 2010.

Ivsic, B., T. Komljenovic, and Z. Sipus, "Performance of uniaxial multilayer cylinders and spheres used for invisible cloak realization," *Proceedings of 5th European Conference on Antennas and Propgation (EUCAP),* pp. 1092–1096, Apr. 2011.

Jafargholi, A. and M. Kamyab, "Pattern optimization in an UWB spiral array antenna," *Progress In Electromagnetics Research M,* vol. 11, pp. 137–151, 2010.

Jiang, Z., J. A. Bossard, and D. H. Werner, "Low loss dual polarized matched zero index metamaterials for microwave applications," *Proceedings of IEEE Antennas and Propagation Society International Symposium,* pp. 1–4, Jun. 2009a.

Jiang, Z., J. A. Bossard and D. H. Werner, "Low loss RF modified fishnet metamaterials with optimized negative, zero and unity refractive index behavior," *Proceedings of IEEE on Antennas and Propagation Society International Symposium,* pp. 1–4, Jun. 2009b.

Jiang, Z. H., J. A. Bossard, X. Wang and D. H. Werner, "Genetic algorithm synthesis of impedance-matched infrared ZIMs with wide FOV using a generalized inversion algorithm," *Proceedings of IEEE Antennas and Propagation Society International Symposium,* pp. 1–4, Jul. 2010a.

Jiang, Z. H., Q. Wu, X. Wang and D. H. Werner, "Flexible wide-angle polarization-insensitive mid-infrared metamaterial absorbers," *Proceedings of IEEE Antennas and Propagation Society International Symposium,* pp. 1–4, Jul. 2010b.

Jiang, Z. H., S. Yun, F. Toor, D. H. Werner and T. S. Mayer, "Experimental demonstration of a conformal optical metamaterial absorber," *Proceedings of IEEE Antennas and Propagation Society International Symposium,* pp. 1812–1815, 2011.

Jin, N., and Y. R. Samii, "Particle swarm optimization of miniaturized quadrature reflection phase structure for low-profile antenna applications," *Proceedings of IEEE Antennas and Propagation Society International Symposium*, vol. 2, pp. 255–258, Jul. 2005.

Kahlout, Y. E., and G. Kiziltas, "Optimally designed microstructures of electromagnetic materials via inverse homogenization," *Proceedings of IEEE on Antennas and Propagation Society International Symposium,* pp. 1–4, Jun. 2009.

Kern, D. J., and D. H. Werner, "A genetic algorithm approach to the design of ultra-thin electromagnetic bandgap absorbers," *Microwave and Optical technology letters,* vol. 38, no. 1, pp. 61–64, Dec. 2003*a*.

Kern, D. J., and D. H. Werner, "The synthesis of metamaterial ferrites for RF applications using electromagnetic bandgap structures," *Proceedings on IEEE Antennas and Propagation Society International Symposium* 2003, vol. 1, pp. 497–500, Jun. 2003*b*.

Kern, D. J., D. H. Werner, M. J. Wilhelm and K. H. Church, "Genetically engineered multiband high-impedance frequency selective surfaces," *Microwave and Optical technology letters,* vol. 38, no. 5, pp. 400–403, Sep. 2003*c*.

Kern, D. H., D. H. Werner and M. Lisovich, "Metaferrites using electromagnetic bandgap structures to synthesize metamaterial ferrites," *IEEE Transactions on Antennas and Propagation*, vol. 53, no. 4, pp. 1382–1389, Apr. 2005.

Kim, D., and J. Yeo, "Dual-band long range passive RFID tag antenna using an AMC ground plane," *Journal of Latex Class Files*, vol. 6, no. 1, pp. 1–8, Jan. 2007.

Kim, T. H., M. Swaminathan, A. Engin and B. J. Yang, "Electromagnetic band gap synthesis using genetic algorithms for mixed signal applications," *IEEE Transactions on Advanced Packaging*, vol. 32, no. 1, pp. 13–25, Feb. 2009.

Kollatou, T. M., A. I. Dimitriadis, N. V. Kantartzis and C. S. Antonopoulos, "A bandwidth-enhanced, ultra-thin, wide-angle metamaterial absorber for EMC applications," *Proceedings of the 10th International Symposium on Electromagnetic Compatibility*, pp. 686–689, Sep. 2011.

Kossiavas, C., and J. L. Dubard, "Synthesis of new artificial magnetic conductors for wideband ultra compact antennas," *The Second European Conference on Antennas and Propagation*, pp. 1–6, Nov. 2007.

Kossiavas, C., A. Zeitler, G. Clementi, C. Migliaccio, R. Staraj and G. Kossiavas, "X-band circularly polarized antenna gain enhancement with metamaterials," *Microwave and Optical Technology Letters*, vol. 53, no. 8, pp. 1911–1915, Aug. 2011.

Kumar, P., and A. K. Singh, "Phase only pattern synthesis for antenna array using genetic algorithm for radar application," *International Journal of Radio and Space Physics*, vol.42, pp. 259–264, 2013.

Kwon, D. H. and D. H. Werner, "Low-index metamaterial designs in the visible spectrum," *Optics Express,* vol. 15, no. 15, pp. 9267–9272, Jul. 2007*a*.

Kwon, D. H., L. Li, J. A. Bossard, M. G. Bray, and D. H. Werner, "Zero index metamaterials with checkerboard structure," *Electronics Letters*, vol. 43, no. 6, Mar. 2007c.

Kwon, D. H., P. L. Werner, and D. H. Werner, "Optical planar chiral metamaterial designs for strong circular dichroism and polarization rotation," *Optics Express*, vol. 16, no.16, pp. 11802–11807, Aug. 2008.

Liang, T., L. Li, J. A. Bossard, D. H. Werner, and T. S. Mayer, "Reconfigurable ultra-thin EBG absorbers using conducting polymers," *Proceedings of IEEE Antennas and Propagation International Symposium*, vol. 2B, pp. 204–207, Jul. 2005.

Lim, S. and H. Ling, "Comparing electrically small folded conical and spherical helix antennas based on a genetic algorithm optimization," *Journal of Electromagnetic Waves and Applications*, vol. 23, pp. 1585–1593, 2009.

Liu, L., S. Matitsine, R. F. Huang, and C. B. Tang, "Electromagnetic smart screen with extended absorption band at microwave frequency," *Metamaterials 5*, pp. 36–41, 2011.

McCulloch, W. S., and W. Pitts, "A logical calculus of the ideas immanent in nervous activity," *Bulletin of Mathematical Biophysics*, vol. 5, pp. 115–133, 1943.

Micheli, D., R. Pastore, C. Apollo, M. Marchetti, G. Gradoni, V. M. Primiani, and F. Moglie, "Broadband electromagnetic absorbers using carbon nanostructure-based composites," *IEEE Transactions on Microwave Theory and Techniques*, vol. 59, no. 10, pp. 2633–2646, Oct. 2011.

Mumcu, G., M. Valerio, K. Sertel and J. L. Volakis, "Applications of the finite element method to designing composite metamaterials," *International Conference on Electromagnetics in Advanced Applications*, 2007, pp. 818–821, Sep. 2007.

Moser, H. O., L. K. Jian, H. S. Chen, M. Bahou, S. M. P. Kalaiselvi, S. Virasawmy, S. M. Maniam, X. X. Cheng, S. P. Heussler, S. B. Mahmood, and B. I. Wu, "All-metal self-supported THz metamaterial the meta foil," *Optics Express*, vol. 17, no. 26, pp. 23914– 23919, 2009.

Ni, X., G. V. Naik, A. V. Kildishev, Y. Barnakov, A. Boltasseva, and V. M. Shalaev, "Effect of metallic and hyperbolic metamaterial surfaces on electric and magnetic dipole emission transitions," *Applied Physics B Lasers and Optics*, vol. 103, pp. 553–558, 2011.

Oraizi, H. and A. Abdolali, "Combination of MLS, GA & CG for the reduction of RCS of multilayered cylindrical structures composed of dispersive metamaterials," *Progress In Electromagnetics Research B*, vol. 3, pp. 227–253, 2008.

Oraizi, H., A. Abdolali, and N. Vaseghi, "Application of double zero metamaterials as radar absorbing materials for the reduction of radar cross section," *Progress In Electromagnetics Research*, vol. 101, pp. 323–337, 2010.

Pattanayak, S., B. Choudhury, and A. Patnaik, "Characterization of planar transmission lines using ANN," *Silver Jublee conference on Communication and VLSI Design, CommV-2009*, Oct. 08-10, Vellore, India.

Pradeep, A., S. Mridula, and P. Mohanan, "Design of an edge-coupled dual-ring split-ring resonator," *IEEE Antennas and Propagation Magazine*, vol. 53, no. 4, pp. 45–54, Aug. 2011.

Qiu, M., M. Yan, and W. Yan, "Metamaterials for space applications," *Department of Microelectronics and Applied Physics, Royal Institute of Technology, Sweden*, pp. 1–16, Jul. 2008.

Radovanovic, J., V. Milanovic, D. Indjin, Z. Ikonic, and P. Harrison, "Charge carrier transport in quantum cascade lasers in strong magnetic field," *Acta Physica Polonica A,* vol. 119, no. 2, pp. 99–102, 2011.

Scarborough, C. P., Q. Wu, D. H. Werner, E. Lier, B. G. Martin, and R. K. Shaw, "A square dual polarization metahorn design," *Proceedings of IEEE Antennas and Propagation International Symposium*, pp. 1065–1068, 2011.

Shen, X., T. J. Cui, J. Zhao, H. F. Ma, W. X. Jiang, and H. Li, "Polarization-independent wide-angle triple-band metamaterial absorber," *Optics Express*, vol. 19, no. 10, pp. 9401–9407, Apr. 2011.

Subramanian, M. S. S., K. V. Siddharth, S. N. Abhinav, V. V. Arthi, K. S. Praveen, R. Jayavarshini, and G. A. S. Sundaram, "Design of dual log-spiral metamaterial resonator for x-band applications," *International Conference on Computing, Communication and Applications,* 2012, pp. 1–6, Feb. 2012.

Tavallaee, A. A. and Y. R. Samii, "A novel strategy for broadband and miniaturized EBG designs hybrid MTL theory and PSO algorithm," *Proceedings of IEEE Antennas and Propagation Society International Symposium,* pp. 161–164, Jun. 2007.

Tonn, D. A. and R. Bansal, "Design of a metamaterial-based linear insulated antenna using a genetic algorithm," *International Journal of RF and Microwave Computer-Aided Engineering*, vol. 19, no. 1, pp. 39–49, Jan. 2009.

Viani, F., M. Salucci, F. Robol, G. Oliveri, and A. Massa, "Design of a UHF RFID/GPS fractal antenna for logistics management," *Journal of Electromagnetic Waves and Applications*, vol. 26, pp. 480–492, 2012.

Vidyalakshmi, M. R. and S. Raghavan, "Comparison of optimization techniques for square split ring resonator," *International Journal of Microwave and Optical Technology,* vol. 5, no. 5, pp. 281–286, Sep. 2010.

Wang, Z., Z. Zhang, S. Qin, L. Wang, and X. Wang, "Theoretical study on wave-absorption properties of a structure with left and right handed materials," *Materials and Design*, vol. 29, no. 9, pp. 1777–17779, Oct. 2008.

Wang, X. and D. H. Werner, "Multiband ultra-thin electromagnetic band-gap and double-sided wideband absorbers based on resistive frequency selective surfaces," *Proceedings of IEEE Antennas and Propagation Society International Symposium, APSURSI '09,* pp. 1–4, Jun. 2009.

Weikai X., L. Shutian, and D. Yangzhang, "Design of structural left-handed material based on topology optimization," *Journal of Wuhan University of Technology-Mater. Sci. Ed.*, vol. 25, no. 2, pp. 282–286, Apr. 2010.

Werner, D. H., D. J. Kern, and M. G. Bray, "Advances in EBG design concepts based on planar FSS structures," *Proceedings of the Loughborough Antennas and Propagation Conference 2005 (invited talk)*, pp. 259–262, Apr. 2005.

Werner, D. H., Z. Bayraktar, F. Namin, T. G. Spence, M. D. Gregory, P. L. Werner, and E. A. Semouchkina, "A novel miniature wideband stacked-patch antenna design using matched impedance magneto-dielectric substrates," *Metamaterials*, pp. 373–375, 2009.

Xin-yuan, L., F. J. Huil, Z. Kuang, H. Jun, and W. Qun, "A compact wideband planar inverted-F antenna (PIFA) loaded with metamaterial," *Proceedings in IEEE Cross Strait Quad-Regional Radio Science and Wireless Technology Conference*, pp. 549–551, Jul. 2011.

Xu, S., X. Cheng, S. Xi, R. Zhang, H. O. Moser, Y. Xu, X. Zhang, and H. Chen, "Low scattering broadband cylindrical invisibility cloak in free space," *arXiv:1108.1204v2 [physics.class-ph]*, pp. 1–17, Aug. 2011.

Yaman, F. and A. E. Yilmaz, "Impacts of genetic algorithm parameters on the solution performance for the uniform circular antenna array pattern synthesis problem," *Journal of Applied Research and Technology*, vol. 8, no.3, pp. 378–394, Dec. 2010.

Yu, Z., Y. Feng, X. Xu, J. Zhao, and T. Jiang, "Optimized cylindrical invisibility cloak with minimum layers of non-magnetic isotropic materials," *Journal of Physics D: Applied Physics*, vol. 44, pp. 185102(1)–185102(6), 2011.

Zhao, Y. X., F. Chen, H. Y. Chen, N. Li, Q. Shen, and L. M. Zhang, "The microstructure design optimization of negative index metamaterials using genetic algorithm," *Electronics Letters*, vol. 22, pp. 95–108, 2011.

4 Bacterial Foraging Optimization For Metamaterial Antennas

The increasing demand for wireless capabilities in modern systems has ushered in the development of compact, high performance antennas. Typically, engineers involved in the design of systems for aerospace applications prefer the use of microstrip, conformal antennas for reduction of drag. However, traditional microstrip antennas have low performance characteristics. Research has shown that the inclusion of metamaterial layers in antenna design can significantly improve its performance. Therefore, the design of metamaterial resonating at the same frequency as the antenna under consideration is crucial to the development of high performance antenna systems. This in fact becomes a time consuming procedure as it requires a systematic variation of structural parameters of the metamaterial while simultaneously observing its performance. In this chapter, an attempt has been made to optimize the procedure for metamaterial design by using *bacterial foraging optimization* (BFO). This soft computing technique will reduce the time taken for obtaining optimized structural parameters and enable rapid design of high performance antenna systems.

4.1 Overview

The usage of a metamaterial layer in an antenna results in a system that shows higher performance—gain enhancement and multi-band operation, and better capability of compact design by reduction of mutual coupling, in the case of antenna arrays. This has led to the application of such systems in wireless communications, especially in the aerospace domain [Lafmajani and Rezaei, 2011]. These systems are often realized by loading a microstrip antenna with a metamaterial as shown in Fig. 4.1.

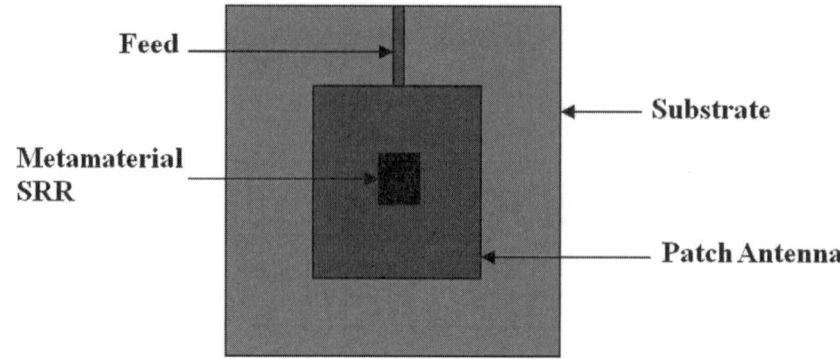

Fig 4.1   Schematic diagram of a metamaterial loaded antenna

As mentioned earlier, the performance of such systems depends on the design of the metamaterial—best performance is observed when the resonant frequency of the metamaterial matches with that of the antenna. Achieving this design objective is a time-consuming task that requires simulation by changing structural parameters iteratively. Efforts are being made to decrease the time involved in obtaining optimized structural parameters using various soft computing techniques such as genetic algorithm, particle swarm optimization, etc.

Genetic algorithm (GA) was used by Kim and Yeo [Kim and Yeo, 2007] to design an AMC (artificial magnetic conductor) for a dual band, passive RFID tag antenna. The algorithm was used to optimize the lumped circuit elements in the equivalent circuit of the AMC. The resultant antenna resonated in the 869.5–869.7 MHz and 910–914 MHz bands, thereby conforming to European and Korean UHF standards, respectively.

A derivate of the genetic algorithm, the microgenetic algorithm was used to optimize a Fabry–Perot cavity antenna that operated within 1.95 GHz–2.5 GHz [Kossiavas and Dubard, 2007]. This frequency range is popularly used in UMTS–WLAN. The Fabry–Perrot based cavity antenna was made up of partially reflecting surface (PRS), perfect electric conductor (PEC) and a patch antenna source. This configuration could be used to simultaneously reduce the thickness and enhance the gain of planar antennas. A similar microgenetic algorithm was used by Kossiavas *et al.* [Kossiavas *et al.*, 2011] to improve the performance of a 2×2 circularly polarized antenna array by optimising a *partially reflective surface* (PRS) metamaterial.

Genetic algorithm can also be used to generate an artificial magnetic metamaterial using a filling element methodology as demonstrated by Chen *et al.* [Chen *et al.*, 2008]. The fitness function used is given in Eq. 4.1 and the algorithm optimized it in order to produce negative magnetic permeability of -1 (C_a =1 and C_b=0 in Eq. 4.1).

$$\text{fitness} = \min\left(\left|\left(\mu'_{eff} + C_a\right)\right| + \left|\left(\mu''_{eff} + C_b\right)\right|\right) \tag{4.1}$$

where, μ'_{eff} is the real part and μ''_{eff} is the imaginary part of the magnetic permeability. Tonn and Bansal [Tonn and Bansal, 2009] used the genetic algorithm to examine the performance of an HF, metamaterial coated, linear insulated antenna. The developed algorithm aims at optimization of antenna gain, bandwidth and gain-bandwidth product. Werner *et al.* [Werner *et al.*, 2009] used the GA in combination with *periodic moment method* (PMM) to find the optimal dielectric constant, material thickness, unit cell size, and FSS geometry of a stacked patched antenna. The aim of the algorithm was to reduce the total antenna thickness. Another soft computing technique called particle swarm optimization integrated with FDTD technique was used by Jin and Samii [Jin and Samii, 2005] in order to determine quadrature refection phase structure topology for miniaturization.

4.2   Challenges in Metamaterial Antenna Design

The biggest challenge while designing metamaterial antennas is to determine the structural parameters of the metamaterial. As shown above, this challenge can be tackled with the use of optimization algorithms for obtaining the structural parameters. The determination of the fitness function for optimization is hence crucial to the efficiency of the algorithm. Further, at times, multiple objectives in the optimization techniques necessitate the usage of multi-objective algorithms, which are complex to implement. Literature has also shown that optimization algorithm is often used along with other analytical techniques such as FDTD, etc. Hence, the final code must be robust and should include implementation of multiple algorithms in a computationally efficient manner.

4.3   BFO for Metamaterial Antenna Design

Bacteria foraging optimization is quite new in the field of microwave engineering. The foraging behaviour of *Escherichia coli* (E. coli) is adopted as an evolutionary computation algorithm. The evolutionary algorithms are based on survival of the fittest concept, where animals with successful foraging strategies eliminate those with poor foraging strategies. As a result, the former are more likely to enjoy reproductive success. Bacterial foraging is a population based search technique and is efficient for problems that require searches over a global space. The foraging behaviour may be further classified into *chemotaxis, swarming, reproduction*, and *elimination and dispersal*. The details of this algorithm are given in Chapter 2 of this book. A brief review of application of bacteria foraging optimization in the field of antenna engineering is given below.

Gollapudi *et al.* [Gollapudi *et al.*, 2009] used BFO in conjunction with PSO for the optimization of resonant frequency of a rectangular microstrip antenna. The computation time was observed to have reduced drastically and the usage of BFO in the development of CAD tools has been presented as a viable option.

Shiju *et al.* [Shiju *et al.*, 2012] used BFO for antenna array pattern synthesis applications. The algorithm was used for beam steering and null replacement in a 10-element linear array with a -30 dB sidelobe level. Similarly, Datta *et al.* [Datta *et al.*, 2008] proposed an improved bacteria foraging algorithm towards null steering in radiation pattern of a six-element antenna array.

Choudhury *et al.* [Choudhury *et al.*, 2013] used bacteria foraging algorithm for fault detection in antenna arrays. The algorithm was tested for both the complete faults and the partial faults in a 22-element antenna array.

4.3.1 Multiband metamaterial fractal antennas

With applications in wireless technology increasing rapidly, most systems use multiple communication links for transfer of information. As a result, miniaturization of antennas, in order to accommodate a maximum number of antennas on the limited surface area of the system in consideration, has become one of the leading areas of research. In this light, the recursive, space-filling characteristic of fractal antennas have made them extremely popular amongst electromagnetic engineers [Werner *et al.*, 1999]. The self-iterative property of fractals offers multiple paths for EM wave propagation and hence these antennas display multiband resonance. Further, the recursive nature of fractal antennas enables implementation of rapid beam forming algorithms and realization of low sidelobe levels [Nur *et al.*, 2011].

Coined by B. B. Mandelbrot (1977), the term fractal is used to denote irregular or fragmented geometries that demonstrate a self-repeating nature, *i.e.* each part is a scaled version of the whole. A result of this self-similarity is a multiband frequency response – a highly desirable property for wireless applications [Nur *et al.*, 2011].

A popular technique used for the design of miniature, highly directive antennas is based on the implementation of space-filling fractals such as Hilbert curves. This was demonstrated by a group of researchers in the design of an inverted F–antenna for usage in wireless sensor networks [Huang *et al.*, 2010]. The design was based on Hilbert geometry and resulted in a 77% decrease in the overall size of the antenna. Apart from using fractals to the design of the antenna itself, studies have also been conducted in order to engineer substrates for patch antennas. In this regard, a substrate containing a multiresonator configuration, with four uncoupled inclusions based on third-order Hilbert structures, has been proposed [Yousefi *et al.*, 2007]. This resulted in a low-loss, dispersion optimized structure, the circuit model of which was also proposed. Further, zero-refractive index metamaterial superstrates have also been designed for patch antennas [Ju *et al.*, 2009]. The addition of this substrate increased the gain of the antenna by 5 dB.

Another fractal geometry, the Koch fractal geometry has been used to decrease the radar cross section (RCS) of a microstrip antenna [Cui *et al.*, 2007]. The addition of this fractal geometry did not affect the radiation pattern of the antenna. A similar reduction in RCS has also been observed upon use of star shaped fractals [Thakare *et al.*, 2010]. The concept of adding fractals to Quasi-Yagi antennas for RFID applications has been explored [Araujo *et al.*, 2011]. The effect of the order and geometrical parameters of a fractal on the performance of a fractal antenna has also been studied by Bengin *et al.* in 2008.

Usage of fractal geometries for improving the performance has also been extended to the field of metamaterials and is well reported in literature. Metamaterials are artificially engineered substances that exhibit unique electromagnetic properties. Palandoken *et al.* [Palandoken *et al.*, 2010] developed a metamaterial based on the Hilbert curves. The engineered metamaterial showed negative permittivity. Numerical analysis for plane wave incidence on a periodic arrangement of the Hilbert curves over a ground plane has also been carried out [McVay *et al.*, 2004]. It is observed that such a configuration forms a high-impedance metamaterial surface.

Improvement in other aspects of performance, viz. increase in gain and bandwidth, decrease in return loss, etc. enables the usage of fractals for a greater degree of applications. One such technique is the inclusion of a metamaterial superstrate layer over the designed fractal antenna. This allows the designer to tune properties, such as permittivity and permeability, in a frequency selective manner. Indeed, this artificial manipulation has enabled researchers to realize negative refractive index, artificial magnetism, negative permittivity, and negative permeability, etc., and has been reported to increase the performance of antennas [Samii *et al.*, 2006].

In this section, a square split-ring resonator with the capability of frequency tuning via control of loaded microsplits is proposed [Ekmekci *et al.*, 2009].

Table 4.1   Characteristic of the fractal patch antenna

Parameters	Resonant frequency of the fractal patch (GHz)					
	3.34	4.9325	6.2975	7.7275	8.7675	12.1150
Return loss (dB)	14.6	16.71	22.1	24.1	10.8	17.5
Gain (dBi)	−0.14	−2.93	3.87	5.67	1.62	4.98
Bandwidth (MHz)	40	85	99.5	220	60	420
Directivity (dBi)	−5.53	3.835	2.802	7.77	5.127	6.93

PSO optimization algorithm is then used to obtain its structural parameters for a desired resonant frequency. The implementation of this algorithm has been discussed in detail in Chapter 2. Based on equivalent circuit analysis, optimum length, width, gap distance, etc., for a particular frequency are obtained.

4.3.1.1 Fractal antenna design

The fractal antenna used in this design is given in Fig. 4.2. The copper antenna occupying an area of 28 mm × 12 mm, was constructed over an FR4 substrate of dimensions 36 mm × 20 mm × 1.6 mm. The dielectric constant of the substrate was taken to be 4.4 [Suganthi *et al.*, 2011]. The antenna was excited using a transmission line of dimension 0.5 mm × 4 mm.

The designed fractal patch antenna was then simulated. It was observed that the antenna resonated at six different frequencies. The performance of the antenna for these frequencies is shown in Table 4.1. From this, the radiation efficiency was computed to be 37.77%. Figures 4.3(a) and 4.3(b) show the return loss and radiation pattern at resonance for the fractal antenna.

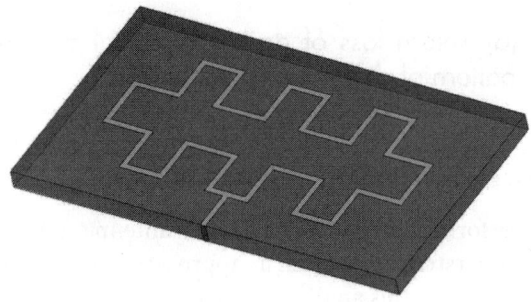

Fig 4.2   Schematic diagram of the fractal antenna

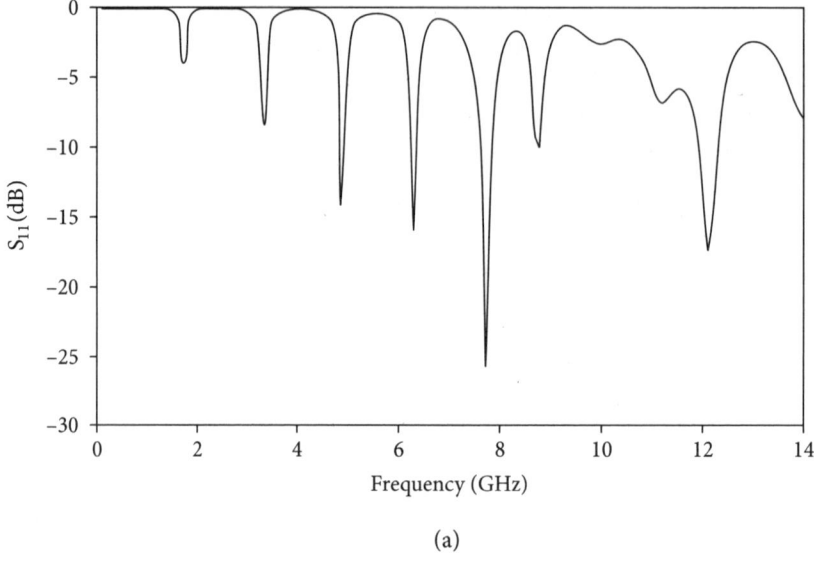

(a)

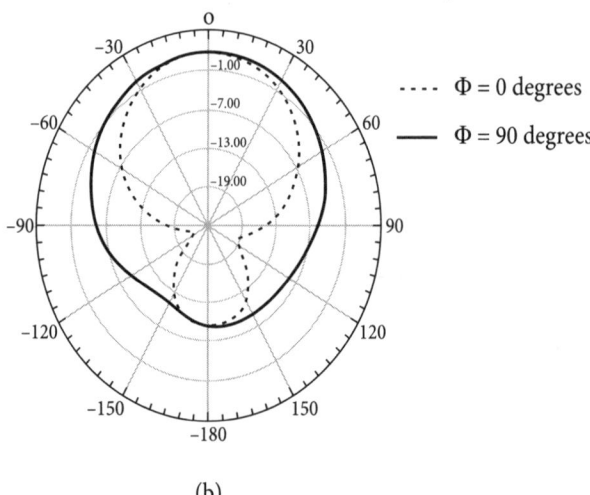

(b)

Fig 4.3 (a) Return loss of designed fractal antenna (b) 2-D radiation pattern of designed fractal antenna

4.3.1.2 Performance enhancement using BFO

As discussed earlier, the performance of the fractal patch antenna can be improved significantly by using metamaterial superstrates. A similar approach has been carried out here. The metamaterial structure chosen for this application is a square split ring resonator (SSRR). This SSRR is printed on a 1.6 mm thick substrate of dielectric constant 4.4. The structure along with its geometrical parameters is shown in Fig. 4.4.

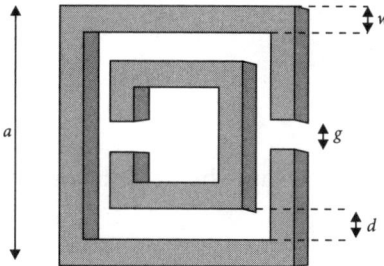

Fig 4.4 Structure of square split ring resonator (structure 1) along with its geometrical parameters where a is the SSRR side length, w is the width of conductor, g is the gap between the rings and d is the distance between the rings

The resonant frequency of the metamaterial is obtained by using equivalent circuit analysis. Using this technique, the SSRR structure can be modelled as an LC tank circuit as shown in Fig. 4.5.

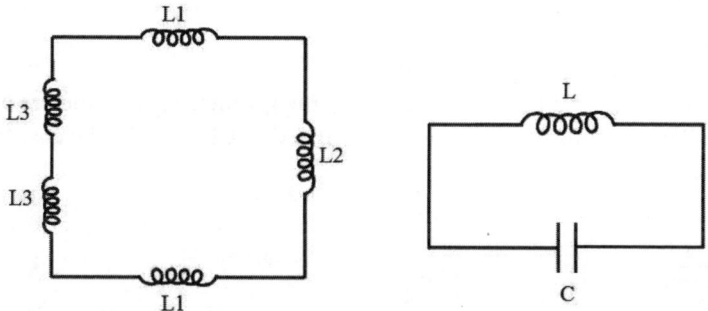

Fig 4.5 Equivalent circuit of square split ring resonator

The resonant frequency is then given by Eq. 4.2

$$f_r = \frac{1}{2\pi\sqrt{LC}}$$
(4.2)

where, L is total inductance and C is gap capacitance which is dependent on structural parameters of SSRR [Billoti *et al.* 2007]. There parameters are given by

$$L = \frac{4.86\mu_0}{2}(a-w-d)\left[\ln\left(\frac{0.98}{\rho}\right)+1.84\rho\right]$$
(4.3)

ρ is the filling factor of the inductance and is given by

$$\rho = \frac{w+d}{a-w-d}$$
(4.4)

The effective capacitance is given by

$$C_s = \left(a - \frac{3}{2}(w + d) \right) C_{pul}$$

(4.5)

where, C_{pul} is the per-unit-length capacitance between the rings which is given as below

$$C_{pul} = \varepsilon_0 \varepsilon_{eff} \frac{\kappa\left(\sqrt{1-k^2}\right)}{\kappa(k)}$$

(4.6)

Here, ε_{eff} is the effective dielectric constant which is expressed as

$$\varepsilon_{eff} = \frac{\varepsilon_r + 1}{2}$$

(4.7)

$K(k)$ denotes the complete elliptical integral of the first kind with k expressed as

$$k = \frac{d}{d + 2w}$$

(4.8)

In order to obtain the desired resonant frequency, the geometrical parameters of the SSRR are tuned using the BFO optimization algorithm discussed in Chapter 2. The cost function of this algorithm is taken as

$$f_{err} = \frac{|f_d - f_c|}{f_d}$$

(4.9)

where, f_c is the frequency arrived at by the equivalent circuit analysis and f_d is the desired frequency. As per the algorithm, the problem specific parameters were assigned as shown in Table 4.2.

Table 4.2 List of parameters used in BFO for design of double ring SSRR

Parameters	Description	Value
P	Dimension of search space	1
S	Total number of bacteria in the population	4
N_c	The number of chemotaxis steps	6
N_s	The number of swimming steps	4
N_{re}	The number of reproduction steps	4
N_{ed}	The Number of elimination and dispersal steps	2
P_{ed}	Elimination-dispersal with probability	0.25
$C(i)$	Tumble step size in the random direction	0.05*ones $(s,1)$

Using this algorithm, the length of side, width of square ring, distance between the rings, gap length, and gap length were found to be 2.8 mm, 0.3 mm, 0.3 mm, and 0.3 mm, respectively. An SSRR was designed using these optimized values and placed exactly in the middle of the fractal patch antenna (Fig. 4.6). This entire assembly was then simulated so as to obtain the S–parameters of the metamaterial. The permittivity and permeability were obtained using parameter retrieval techniques.

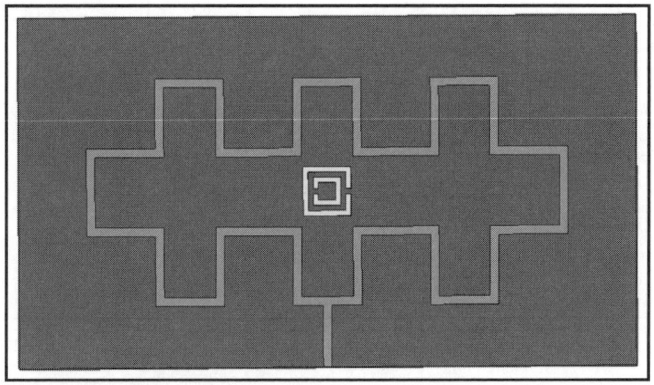

Fig 4.6 Schematic diagram of fractal antenna with metamaterial structure 1

Another SSRR was also designed using the same structural parameters. However, in this case, micro-splits were included on the outer ring of the SSRR (Fig. 4.7). The structure, which will be referred to in this text as Structure 2 (the previous being Structure 1 as shown in Fig. 4.4), was then placed in the same position as Structure 1 and simulated. The results of the simulation are given in Fig. 4.8. The micro-splits in Structure 2 help it resonate at multiple frequencies thereby improving the performance in multiple bands.

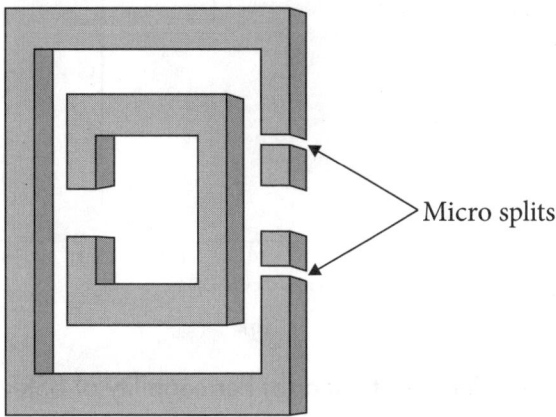

Fig 4.7 SSRR with micro-splits (considered as Structure 2)

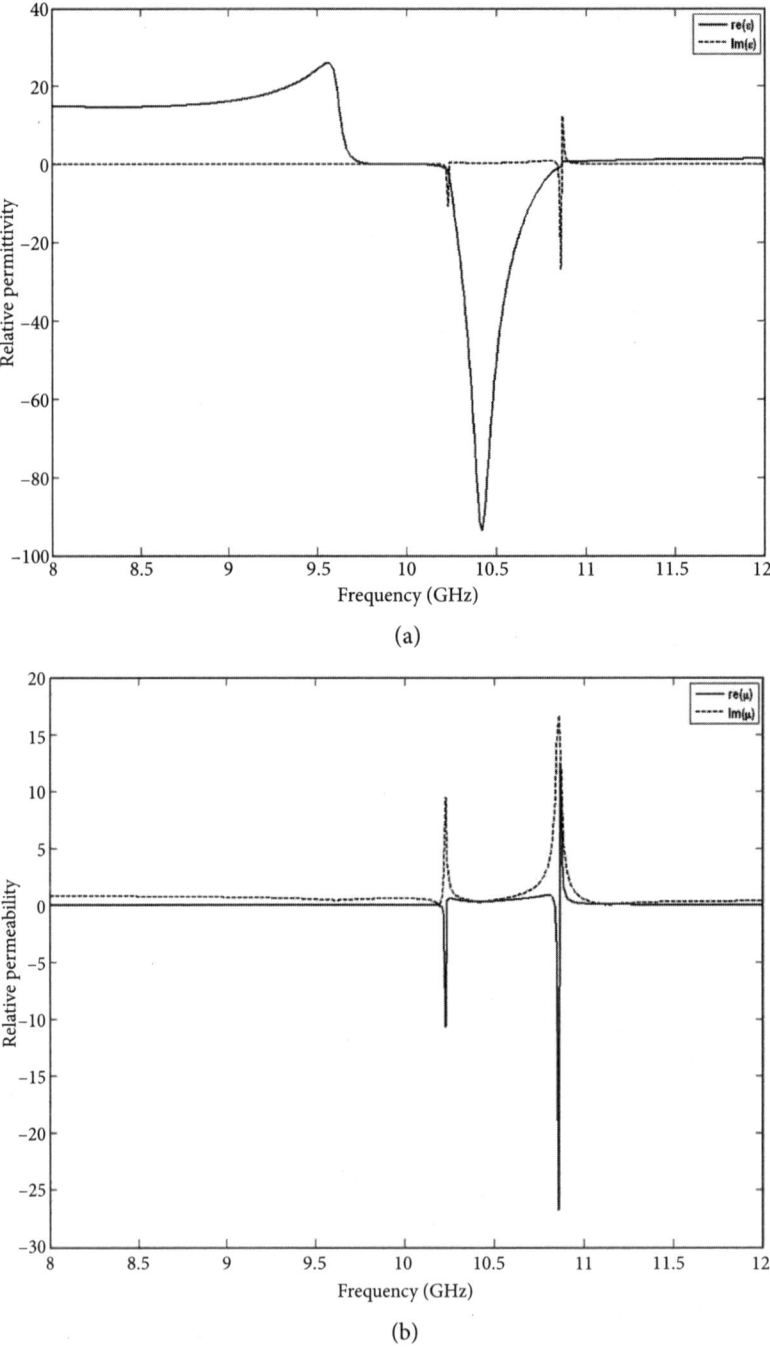

Fig 4.8 (a) Permittivity and (b) Permeability of SSRR with micro splits

It may be observed that though the addition of a metamaterial to the fractal patch antenna improves the performance, the latter is not consistent over all the frequencies. Therefore, a

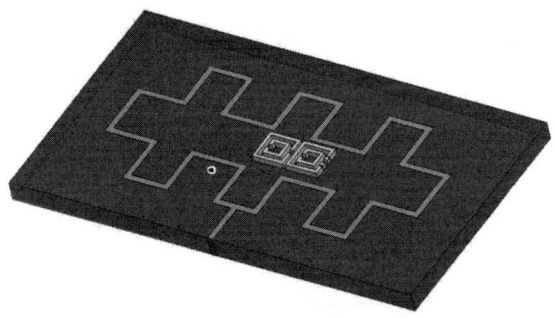

Fig. 4.9 Fractal antenna with two types of metamaterial SSRR (Structure 1 and Structure 2)

Table 4.3 Characteristics of the fractal patch antenna after the addition of metamaterial structure

Parameters	Resonant frequency of the fractal patch (GHz)					
	3.37	4.96	6.42	7.76	8.93	12.37
Bandwidth (MHz) (without SRR)	40	85	99.5	220	60	420
Bandwidth (MHz) (with SRR)	46	340	120	220	120	460
Return loss (dB) (with SRR)	12.50	31.52	26.8	20.35	13.30	21.26
Directivity (dBi) (with SRR)	−3.14	4.86	3.31	7.79	5.69	6.21

combination of Structure 1 and Structure 2 is proposed in order to obtain better performance over a wider bandwidth (Fig. 4.9). Considerable improvement in performance in terms of VSWR, return loss, gain, multi-band characteristics etc. is observed using this configuration of two SSRRs (Figs. 4.10(a) and 4.10(b)). A comparison between the bandwidth of the fractal patch antenna with and without the metamaterial layer as well as performance with the metamaterial layer is given in Table 4.3.

The radiation efficiency is found to increase by 40% with the metamaterial layer. Further, the directivity of the antenna improves and the return loss decreases at the resonant frequency. Figures 4.10 (a) and (b) show the return loss and the radiation pattern of the fractal patch antenna with metamaterial array.

This example demonstrates the effectiveness of BFO as an optimization algorithm for metamaterial design. The metamaterial designed using this algorithm is used as a superstrate for a fractal patch antenna. Significant performance improvements are observed after the addition of the metamaterial array.

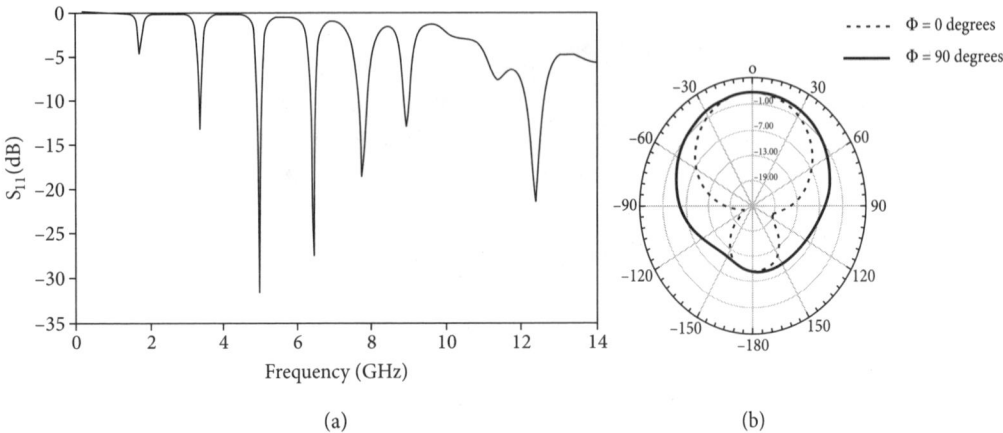

(a)   (b)

Fig 4.10 (a) S_{11} of fractal antenna with metamaterial structures as superstrate (b) 2-D radiation pattern of the fractal antenna with metamaterials as superstrate (Fig. 4.9)

4.3.2 Mutual coupling reduction

The increase in the demand for compact antennas for wireless systems has been discussed in the previous section. This trend has been primarily due to an increase in the number of wireless functionalities in the system, typically involving transmission and reception at multiple frequencies. Therefore a large number of antennas are placed very close together. This results in mutual coupling, which degrades the performance of the antennas [Fletcher *et al.*, 2003]. This effect is extremely predominant in elements of microstrip patch antenna arrays. A popular technique to combat the effects of mutual coupling involves the use of *electronic band gap* (EBG) structures [Rajo–Iglesias *et al.*, 2008]. However, this technique results in very complicated designs.

In order to avoid this complexity, a technique of reducing the coupling by using metamaterials is discussed in this section. The proposed metamaterial is made up of square split ring resonators (SSRR). Optimization of the metamaterial structure is carried out using BFO. Reduction in mutual coupling between two microstrip patch antennas operating at 2.36 GHz using the designed metamaterial is demonstrated through simulation.

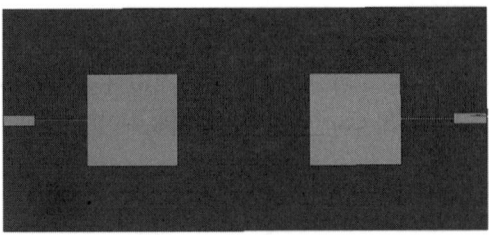

Fig 4.11 Schematic diagram showing design of antenna array

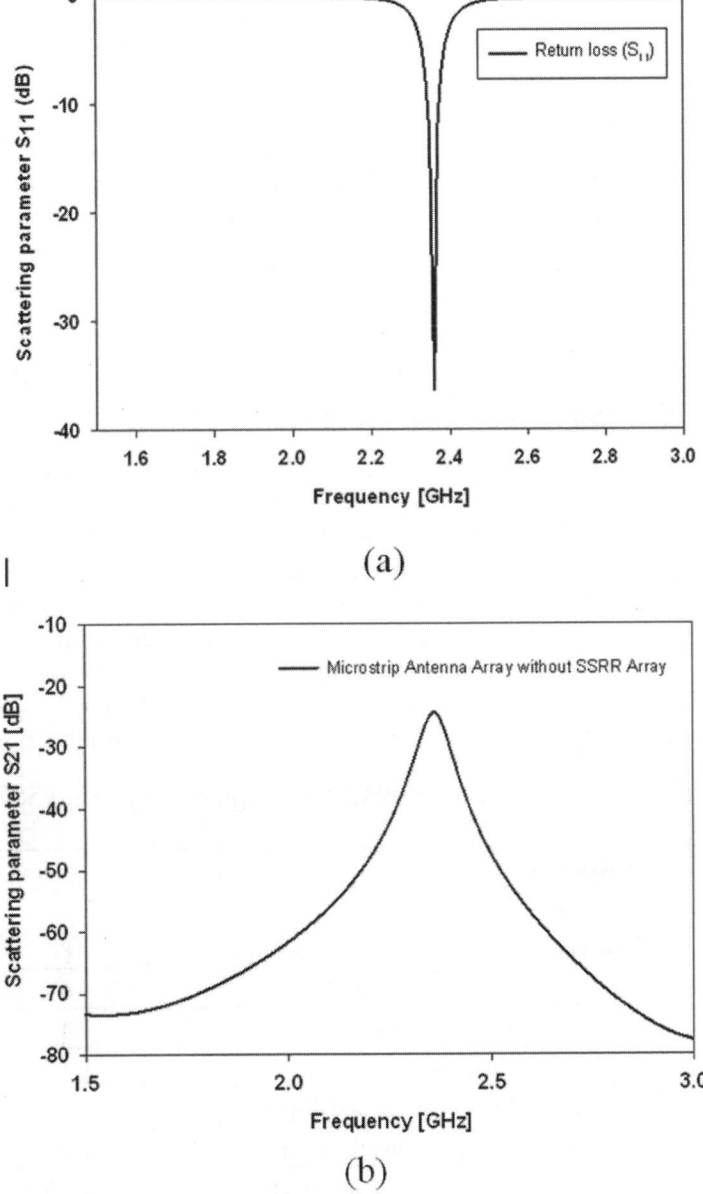

Fig 4.12   (a) Return loss in antenna element (b) Mutual coupling between the two patch antennas

4.3.2.1 Design of microstrip antenna array

A two-element microstrip patch antenna array is designed in this section. The frequency of operation is chosen to be 2.36 GHz in order to design an antenna that can be used in the ISM band. Two patches of dimensions 41.08 mm × 41.08 mm are placed on a substrate made up of Duroid (220 mm × 98.5 mm × 1.57mm) as shown in Fig. 4.11. The dielectric constant of the substrate is taken to be 2.2. Impedance matching of the feed is ensured by choosing proper dimensions for the feed network.

The two patches are placed at a distance 59.74 mm. It should be noted here that this distance is less than $\lambda/2$. Therefore, the effects of mutual coupling between the antenna elements will be very strong. However, effective reduction in mutual coupling can be shown only in such scenarios. In fact, the success of this technique will bring out the possibility of fabrication of highly compact arrays due to the removal of the constraint on the minimum distance between elements. The design is then simulated and the S-parameter of each element and mutual coupling between the two elements are plotted as shown in Figs. 4.12 (a) and (b). Figure 4.12(b) shows a maximum coupling of -24 dB.

4.3.2.2 Mutual coupling reduction using metamaterial

Using the BFO algorithm given in Chapter 2, a square split ring resonator (SSRR) is designed for 2.36 GHz. The procedure for the design of this SSRR metamaterial is same as the one described in Section 4.4.1.3. After optimization using BFO, the length of the SSRR, the width of the rings, the distance between the rings and the gap length are found to be 4.90 mm, 0.21mm, 0.3 mm and 0.3 mm, respectively, by using a double ring SSRR. Further, a single ring SSRR has been optimized using BFO towards easy fabrication and to avoid design complexity. The extracted length l, width w, and the gap spacing g, are 9 mm, 0.18 mm and 0.125 mm, respectively. The BFO parameters considered in this case are given in Table 4.4. The designed unit cell is simulated and its scattering parameters, permittivity, and permeability are shown in Fig. 4.13 (a) and (b).

Table 4.4 List of parameters used in BFO for design of single ring SSRR

Parameters	Description	Value
p	Dimension of search space	1
s	Total number of bacteria in the population	8
N_c	The number of chemotaxis steps	6
N_s	The number of swimming steps	4
N_{re}	The number of reproduction steps	4
N_{ed}	The Number of elimination and dispersal steps	2
P_{ed}	Elimination-dispersal with probability	0.25
$C(i)$	Tumble step size in the random direction	0.05*ones(s,1)

This optimized SSRR structure is then arranged to create a metamaterial array, which is placed in between the two elements of the antenna array. The position of the array is varied iteratively so as to find the exact position that results in minimum mutual coupling. Four positions (with respect to the element in the left in the Fig. 4.11) are chosen for this study and are shown in Fig. 4.14. At Position 1, SSRR array placed at a distance of 3.9 mm (0.0307 λ), the mutual coupling is seen to reduce by 2.1 dB (Fig. 4.15). At Position 2, distance of 14.9 mm (0.1172 λ), reduction in mutual coupling is found to be 1.1 dB. Due to symmetry, Positions 3 and 4 show the same results as 2 and 1, respectively. Hence, control over mutual coupling can be obtained by varying the position of the SSRR with respect to the patch.

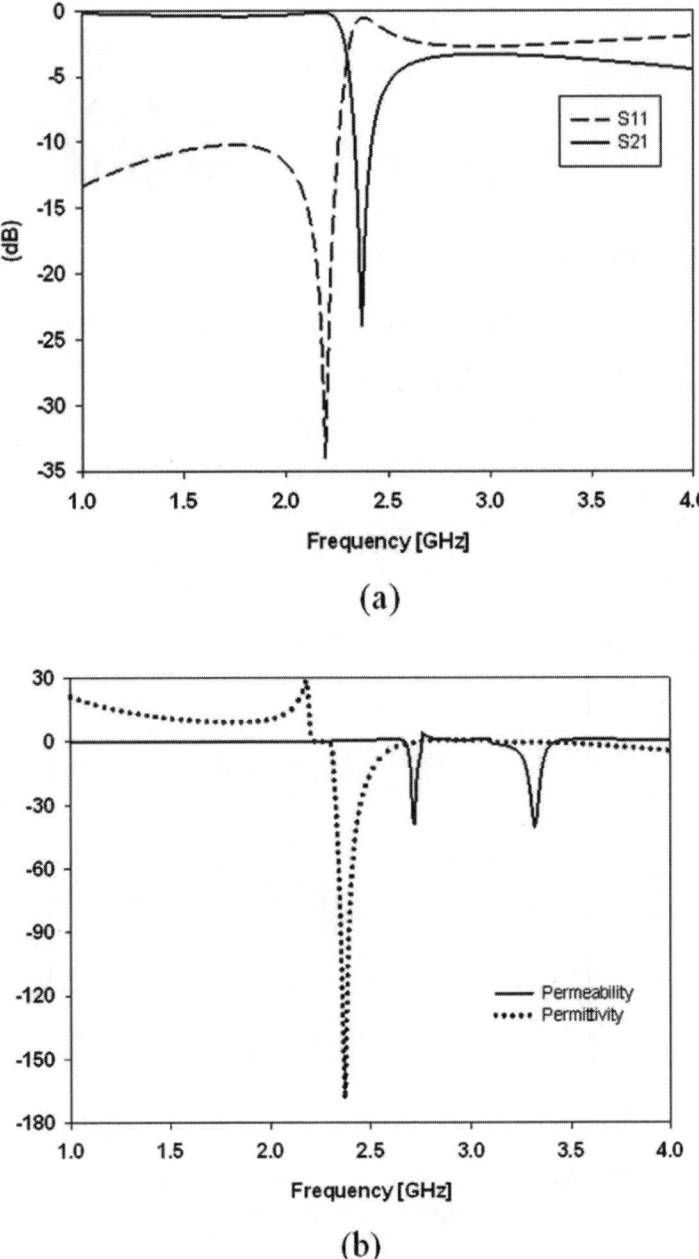

(a)

(b)

Fig 4.13 (a) Return loss of metamaterial unit cell, SSRR (b) Permeability and permittivity of SSRR

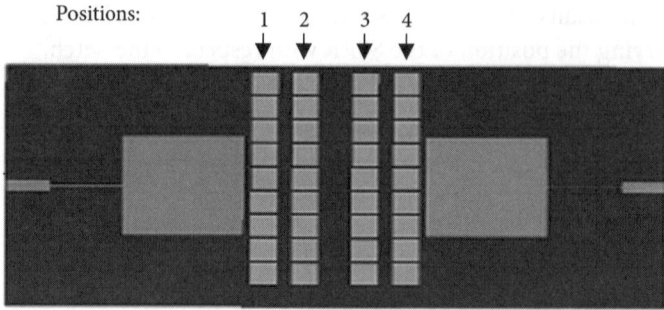

Fig. 4.14 Postions of metamaterial array with respect to antenna array

Without the metamaterial array, the maximum coupling between the two elements of the patch was found to be -24.4 dB. When the SSRR array was placed at Position 1 (or Position 4), the mutual coupling reduced to -26.5 dB. At Position 2 (or position 3), the mutual coupling is observed to be -25.3 dB. Best reduction in coupling was observed in positions 1 and 3 with a decrease by 2.1 dB.

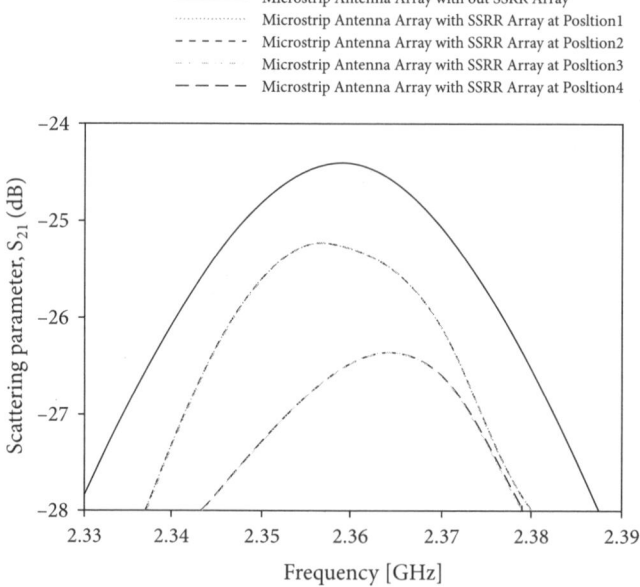

Fig 4.15 Mutual coupling between the two elements of the antenna array for different positions

Reduction in mutual coupling will lead to better performance of compact arrays. The distance between elements in an antenna array can be reduced to below $\lambda/2$ by placing a metamaterial array in between the elements. Design of the metamaterial is crucial to the performance of the system due to the fact that SSRR metamaterials are frequency selective. The task of finding the optimum structural parameters for the SSRR for a desired frequency is made simple by implementing a BFO algorithm as shown above. This example along with the example on improvement of performance of fractal patch antennas demonstrate that BFO yields accurate results and can be used as a tool for metamaterial design.

4.4 Summary

In this chapter, the implementation of *bacterial foraging optimization* (BFO) algorithm for metamaterial design has been discussed. The designed metamaterial has been used to improve the performance of fractal antennas as well as to reduce the mutual coupling in antenna arrays. These examples demonstrate the usage of BFO as a tool for rapid design of metamaterial. It is apt to be noted here that the examples illustrated in this chapter pertain to single-objective optimization, where as practical scenarios often require multi-objective optimization. In such situations, the soft computing algorithm must accommodate multiple objectives, which is described in Chapter 8.

References

Araujo, H. X., D. S. E. Barbin, and L. C. Kretly, "Design of UHF Quasi-Yagi antenna with metamaterial structures for RFID applications," *Microwave and Optoelectronics Conference, IEEE*, pp. 8–11, Nov. 2011.

Bengin, V. C., V. Radonic, and B. Jokanovic, "Fractal geometries of complementary split-ring resonators," *IEEE Transactions on Microwave Theory and Techniques*, vol. 56, no. 10, Oct. 2008.

Bilotti, F., A. Toscano, and L. Vegni, "Design of spiral and multiple split-ring resonators for the realization of miniaturized metamaterial samples," *IEEE Transactions on Antennas and Propagation*, vol. 55, no. 8, pp. 2258–2267, Aug. 2007.

Chen, P. Y., C. H. Chen, H. Wang, J. H. Tsai, and W. X. Ni, "Synthesis design of artificial magnetic metamaterials using a genetic algorithm," *Optics Express*, vol. 16, no.17, pp. 12806–12818, Aug. 2008.

Choudhury, B., O. P. Acharya, and A. Patnaik, "Bacteria foraging optimisation in antenna engineering: An application to array fault finding," *International Journal of RF and Microwave Computer-Aided Engineering*, vol. 23, no. 2, pp. 141–148, Mar. 2013.

Cui, G., Y. Liu, and S. Gong, "A novel fractal patch antenna with low RCS," *Journal of Electromagnetics Waves and Application*, vol. 21, no. 15, pp. 2403–2411, 2007.

Datta, T., I. S. Mishra, B. B. Mangaraj, and S. Imtiaj, "Improved adaptive bacteria foraging algorithm in optimisation of antenna array for faster convergence," *Progress in Electromagnetic Research C*, vol. 1, pp. 143–157, 2008.

Ekmekci, E., K. Topalli, T. Akin, and G. Turhan-Sayan, "A tunable multi-band metamaterial design using micro-split SRR structures," *Optics Express*, vol. 17, no. 18, pp. 16046–16058, Aug. 2009.

Fletcher, P. N., M. Dean, and A. R. Nix, "Mutual coupling in multi-element array antennas and its influence on MIMO channel capacity," *Electronic Letters,* vol. 39, no. 4, pp. 342–344, Feb. 2003.

Gollapudi, S. V. R. S., S. S. Pattnaik, O. P. Bajapai, S. Devi, K. M. Bakwad, and P. K. Pradyumna, "Intelligent bacterial foraging optimisation technique to calculate resonant frequency of RMA," *International Journal of Microwave and Optical Technology,* vol. 4, pp. 67–75, Mar. 2009.

Huang, J. T., J. H. Shiao, and J. M. Wu, "A miniaturized Hilbert inverted-F antenna for wireless sensor network applications," *IEEE Transactions on Antennas and Propagation,* vol. 58, no. 9, pp. 3100–3103, Sep. 2010.

Jin, N. and Y. R. Samii, "Particle swarm optimisation of miniaturized quadrature reflection phase structure for low-profile antenna applications," *IEEE Antennas and Propagation Society International Symposium,* vol. 2, pp. 255–258, Jul. 2005.

Ju, J., D. Kim, W. J. Lee, and J. I. Choi, "Wideband high-gain antenna using metamaterial superstrate with the zero refractive index," *Microwave and Optical Technology Letters,* vol. 51, no. 8, pp. 1973–1976, Aug. 2009.

Kim, D. and J. Yeo, "Dual-band long range passive RFID tag antenna using an AMC ground plane," *IEEE Transactions on Antennas and Propagation,* vol. 60, no. 6, pp. 2620–2626, June 2007.

Kossiavas, C., A. Zeitler, G. Clementi, C. Migliaccio, R. Staraj, and G. Kossiavas, "X-band circularly polarized antenna gain enhancement with metamaterials," *Microwave and Optical Technology Letters,* vol. 53, no. 8, pp. 1911–1915, Aug. 2011.

Kossiavas, C. and J. L. Dubard, "Synthesis of new artificial magnetic conductors for wideband ultra compact antennas," *The Second European Conference on Antennas and Propagation,* pp. 1–6, Nov. 2007.

Lafmajani, I. A. and P. Rezaei, "Miniaturized rectangular patch antenna loaded with spiral/wires metamaterial," *European Journal of Scientific Research,* vol. 65, no. 1, pp. 121–130, 2011.

Mandelbrot, B. B., *Les, Objects Fractals – Forme, Hasard et Diemnsions,* 4th Ed., Champs Flammarion, Paris, France, ISBN 2-08-081301, 1995. Translation in English *Fractals. Form, Chance and Dimension,* W.H. Freeman & Co Springer, Netherlands, ISBN: 0716704730, 1977.

McVay, J., N. Engheta, and A. Hoorfar, "High impedance metamaterial surfaces using Hilbert-curve inclusions," *IEEE Microwave and Wireless Components Letters,* vol. 14, no. 3, pp. 130–132, Mar. 2004.

Nur, T. E., S. K. Ray, D. Paul, and T. Mollick, "Design of fractal antenna for ultra-wideband applications," *International Journal of Research and Reviews in Wireless Communications,* vol. 1, no. 3, pp. 66–74, 2011.

Palandoken, M., and H. Henke, "Fractal negative-epsilon metamaterial," *Antenna Technology (iWAT) IEEE,* pp. 1–4, Mar. 2010.

Rajo-Iglesias, E., O. Quevedo-Teruel, and L. Inclan-Sanchez, "Mutual coupling reduction in patch antenna arrays by using a planar EBG structure and a multilayer dielectric substrate," *IEEE Transactions on Antennas and Propagation,* vol. 56, no. 6, pp. 1648–1655, June 2008.

Samii, Y. R., "Metamaterials in antenna applications: Classifications, designs and applications," *Proceedings of IEEE International Workshop on Antenna Technology, Small Antennas and Novel Metamaterials*, pp. 1–4, Mar. 2006.

Shiju, R. M. and N. Venkateswaran, " Optimisation of linear array antenna pattern synthesis using bacterial foraging algorithm," *Proceedings of International Conference on Recent Advances in Computing and Software Systems*, pp. 130–134, Apr. 2012.

Suganthi, S., S. Raghavan, and D. Kumar, "Miniature fractal antenna design and simulation for Wireless Applications," *International Conference on IEEE Recent Advances in Intelligent Computational Systems (RAICS2011) Trivandrum*, pp. 51, Sep. 2011.

Thakare, Y. B. and Rajkumar, "Design of fractal patch antenna for size and radar cross-section reduction," *IET Microwaves Antennas Propagation*, vol. 4, no. 2, pp. 175–181, 2010.

Tonn, D. A. and R. Bansal, "Design of a metamaterial-based linear insulated antenna using a genetic algorithm," *International Journal of RF and Microwave Computer-Aided Engineering*, vol. 19, no. 1, pp. 39–49, Jan. 2009.

Werner, D. H., Z. Bayraktar, F. Namin, T. G. Spence, M. D. Gregory, P. L. Werner, and E. A. Semouchkina, "A novel miniature wideband stacked-patch antenna design using matched impedance magneto-dielectric substrates," *Metamaterials*, pp. 373–375, 2009.

Werner, D. H., R. L. Haupt, and P. L. Werner, "Fractal antenna engineering: The theory and design of fractal antenna arrays," *IEEE Antennas and Propagation Magazine*, vol. 41, no. 5, pp. 37–58, 1999.

Yousefi L., and O. M. Ramahi, "Miniaturized wideband antenna using engineered magnetic materials with multi-mesonator inclusions," *IEEE International Symposium on Antennas and Propagation Society*, 2007.

PSO for Radar Absorbers

As the name suggests, absorbers are devices that absorb electromagnetic radiation incident on them. Absorbers are hence used in applications where minimum reflection is desired such as construction of anechoic chambers, stealth aircraft, etc. Absorbers are also used to enhance the performance of detectors in various imaging systems like terahertz spectroscopy. Absorbers generally comprise of layers of different material placed one behind the other. Due to the nature of its construction, absorbers are extremely band specific. The selection of the material parameters and thickness of these layers determines the frequency and bandwidth of operation. This selection process is complex and time consuming as the designer must focus on the combination of material as well as its thickness simultaneously.

In this chapter, particle swarm optimization (PSO) is used to optimize the absorbers in a time efficient manner. First, the implementation of PSO for optimising conventional microwave absorbers is discussed. Following this, PSO based optimization of a metamaterial terahertz absorber for biomedical applications is presented.

5.1 Introduction

An electromagnetic absorber is a structure that ideally absorbs all the incident electromagnetic radiation without any transmission or reflection. This is achieved by selecting materials of specific dimensions. Often, designs employ the arrangement of multiple layers of varying dimensions in order to achieve maximum absorption. At the same time, applications in stealth technology imposes another constraint on the design namely that of thickness. These two design parameters conflict each other and the designer is forced to arrive at a trade-off between the two. As mentioned previously, this task is time-consuming. As a result, researchers have turned towards soft-computing in order to design optimized RAM structures.

Abundant literature is available for implementation of genetic algorithm and micro-genetic algorithm for RAM optimization. Chakravarty *et al.* [Chakravarty *et al.*, 2001] used the same in order to design an FSS based broadband microwave absorber. The work also shows that implementation of micro-genetic algorithm over genetic algorithm considerably speeds up the computation time. The algorithm was designed to simultaneously select the best materials and their thicknesses as well as vary the structural parameters of the FSS for optimized performance.

A novel idea for the fabrication of ultrathin absorbers using electromagnetic band gap materials was presented by Kern *et al.* [Kern *et al.*, 2003]. The technique involved replacing previously known FSS-resistive sheet designs with a lossy, high impedance FSS layer. A micro-genetic algorithm was used to optimize this FSS layer.

Weile *et al.* [Weile *et al.*, 1996] implemented different pareto-front based genetic algorithms in order to achieve a trade-off between thickness and reflection in microwave RAM designs. These algorithms considered frequency of operation and incidence angles of EM waves in order to design an appropriate RAM from a given database of materials. It was observed that non-dominated sorting genetic algorithm (NSGA) provided the best results.

Michielssen *et al.* [Michielssen *et al.*, 1993] discussed the implementation of combinatorial genetic algorithm for the realization of high performance, physically realisable RAM designs. The total number of layers was pre-defined and materials were selected from a given database. The algorithm was probability based, thereby increasing the speed with which the global maximum is searched.

Particle swarm optimization has also been used for design of RAM structures. Examples of this implementation are discussed later in Section 5.4.

5.2 Types of Radar Absorbers

Depending on the bandwidth of absorption, absorbers can be broadly classified into two different categories: narrowband absorbers and broadband absorbers [Vinoy and Jha, 1996].

Narrowband absorbers exploit the properties of quarter wavelength transmission lines in order to eliminate the imaginary part of their surface impedance. Salisbury screens, Dallenbach layers, magnetic absorbers, and CA–RAM are examples of narrowband absorbers. Typically, these designs involve the use of a single resistive layer and a spacer.

On the other hand, broadband absorbers use multiple layers and other inherent properties of materials like chirality, etc., in order increase absorption. Examples of these kinds of absorbers include Jaumann absorber, inhomogeneous absorber, geometric transition absorber, and chiral absorber.

5.2.1 Salisbury screen

Salisbury screens are one of the simplest absorbers that can be fabricated. They are made up of resistive sheets placed at a suitable distance from a metallic plane. This distance is usually given as $\lambda/4$ and this transforms the zero impedance on the metal surface to an open circuit. The thickness of the sheet affects its conductivity and must be chosen appropriately. Furthermore, these sheets show either very high electric loss tangent or magnetic loss tangent. When the magnetic loss tangent is large, Salisbury screens are also called magnetic absorbers. In a practical scenario, the gap between the resistive sheet and the metallic layer is filled with materials, which mimic the properties of free-space. Additionally, research has been carried out to overcome the limitation of the distance at which the resistive sheet must be placed. This involves the introduction of capacitively loaded resistive sheets.

5.2.2 Magnetic absorbers

Magnetic absorbers are a variation of Salisbury screens. In this case, a highly lossy magnetic material is placed on the metal sheet directly. Losses are generally due to presence of ferrites or carbonyl ions in the resistive sheet. Magnetic absorbers enable the construction of light-weight RAMs that can be easily mounted on metallic structures. This property makes them more popular than the Salisbury screen.

5.2.3 Dallenbach layer

Dallenbach layers are high permittivity dielectric sheets that can be placed over metallic surfaces. Unlike Salisbury screens that are analyzed either for electric or magnetic properties, Dallenbach layers are analyzed by taking both permittivity and permeability into account.

5.2.4 Circuit analog RAM

Circuit analog RAMs are a popular form of narrowband RAMs. They are constructed by etching metallic patterns on the surface of a dielectric layer. The resonant frequencies of these structures ensure the dissipation of any EM energy incident on the absorber. The patterns are generally associated with a characteristic reactance. As a result, they can be easily represented in the form of a circuit containing lumped elements, resulting in the name circuit analog RAM.

5.2.5 Jaumann absorber

Jaumann absorber designs are broadband absorbers. Typically, Jaumann absorbers consist of multiple resistive sheets separated with dielectric spacers. The bandwidth of operation can be controlled by increasing the number of layers in the absorber design and selecting the correct values of permittivity for the dielectric spacers. The thickness of both the layers viz. resistive sheet and dielectric spacer can be varied in order to obtain minimum reflection.

5.3   Radar Absorber Design Procedure

Two concepts are used in the design of radar absorbing material: the theory of matched characteristic impedance, where the front face characteristic impedance of an absorber is equated to that of free space, and the theory of matched wave impedance, where the input impedance of the front face is made equal to the impedance of free space. The former serves as a condition for zero-reflection while the latter ensures that entire energy is transmitted into the absorber [Vinoy and Jha, 1996].

In order to simultaneously minimize the reflection and transmission through a material, careful selection of the permittivity (ε) and permeability (μ) of the material is required. In order to minimize the reflection, the impedance of the RAM should be matched to that of free space. The impedance of the RAM, z is given by

$$z = \sqrt{\mu / \varepsilon} \qquad\qquad (5.1)$$

Therefore, for perfect match, the ratio of relative permeability to relative permittivity becomes unity.

The imaginary parts of permittivity and permeability correspond to the losses in the material. In order to minimize transmission, the imaginary parts of ε and µ should be as large as possible. Practically, this is achieved by placing a metal layer that acts a ground with thickness greater than the skin depth of the incident wave.

The relationship between the transmittance, T and reflectance, R is then given as

$$\left|T^2\right| + \left|R^2\right| + A = 1 \tag{5.2}$$

where, A denotes the absorption in the metamaterials layer. When $T=R=0$, the absorption becomes unity.

5.4   PSO for Design Optimization

Proposed by Kennedy and Eberhart in 1995, particle swarm optimization (PSO) is an optimization technique used in multidimensional discontinuous problems [Choudhury *et al.*, 2012]. This technique mimics the behaviour of swarm intelligence and movements; it is analogous to the behaviour of bees in a field with a goal of finding the location with the highest density of flowers. In this chapter, the following sections describe the implementation of PSO along with the algorithm [Robinson and Samii, 2004]. It should be noted that in these examples, the resonant frequency is considered to be an *a priori* requirement.

Instances of implementation of genetic algorithm in RAM have been discussed in Section 5.1. Other algorithms like the particle swarm optimization have also been used for the same purpose; PSO is simpler and executes faster than genetic algorithms. In this section, we provide a detailed description of the same, following which the concept of PSO will be applied in order to optimize certain RAM designs.

The concept of using particular swarm optimization for RAM design optimization is not new. In fact, researchers have demonstrated the implementation of modified versions of the same [Liu *et al.* 2009] for designing RAMs. This improved PSO was faster and at the same time prevented the algorithm from falling into local minima. This was done by monitoring the changes in iterations and changing the strategy used if the result remained invariant for a specific time-frame.

Others have used multi-objective pareto-front based algorithms to achieve their goal, thereby arriving at multiple solutions. Chamaani *et al.* [Chamaani *et al.*, 2007] used a modified MOPSO technique in order to optimize a 5 layered RAM.

Cui *et al.* [Cui *et al.*, 2005] proposed the implementation of a parallel PSO for RAM design. The idea was to calculate the fitness function of each agent in parallel and update the global best after the fitness of all the agents were calculated. This type of PSO is called synchronous PSO. In contrast, conventional or asynchronous PSO updates the global best after the fitness of each agent is calculated. While the basic structure of the algorithm remained similar to that of asynchronous PSO, a special operator had to be introduced to prevent the algorithm from converging to the boundary sub-optimally. This technique was found to be faster than asynchronous PSO and GA in certain problem types.

5.4.1 Jaumann absorber optimization

The optimization methodology of RAM design can be of two types, namely optimization of passive absorber where the thickness and materials are fixed but the arrangement in the layers can be optimized to achieve the optimum absorption; or an active absorber where the thickness and material parameters can be changed for achieving a pre-specified absorption and desired band of frequency. The objective functions thus vary according to the desired specifications. In this section, an example of optimization technique applied to a Jaumann absorber is presented.

As mentioned in the earlier sections, Jaumann absorber consists of resistive sheets sandwiched between dielectric spacers. The properties of these layers, viz. conductivity in case of resistive sheets and permittivity in case of dielectric layers are different for different layers. In this section a 14-layer absorber consisting of 7 resistive sheets and 7 dielectrics is optimized for wide-band performance. Let us assume that the application requires maximum of -20dB reflectivity over a frequency range of 2–10 GHz. The performance of the absorber design depends on permittivity of spacer layers, their thicknesses as well as resistivity and thickness of the resistive sheets. In order to simplify the problem, only the thicknesses of the dielectric spacers are varied with all other parameters kept constant. The relative permittivity of all the spacers is taken to be 1.03. The resistive sheets of resistivity (in ohm/square) 300, 650, 1400, 3000, 6500, 14000 and 30000 of thickness 0.01cm are chosen. The total reflection was calculated by computing the wave impedances of each layer iteratively [Vinoy and Jha, 1996].

As the design goal is to achieve -20 dB reflectivity over the entire range of 2–10 GHz, one should go for an optimization technique that will take the thickness of each dielectric spacer as input and the reflectivity as output. In this work, PSO has been selected as the optimization technique due to its merit as a fast converging algorithm. The PSO algorithm explained in Chapter 2, Section 2.3 is used for optimization of the fitness function given below:

$$f = \max [20 \log_{10} (\text{reflection})] \tag{5.3}$$

For each particle in the simulation, the reflection at each frequency over the entire frequency range is calculated. Since the objective is to minimize the reflection over this band, the fitness function corresponds to the maximum value amongst the calculated values of reflection.

The developed PSO algorithm was implemented with the parameters given in Table 5.1.

Table 5.1 Parameters considered for PSO algorithm for RAM design

PSO Parameters	Value	Definition
W	1–0.4	Inertial weight decreases from 1 to 0.4
$c1$	0.5	Cognitive Parameter
$c2$	0.2	Social Parameter
Np	500	Number of particles
Nd	7	Number of dimensions
Nt	1500	Number of time steps
X_{min}	0.5	Scalar, min. for particle position
X_{max}	3	Scalar, max. for particle position
V_{min}	−5	Scalar, min. for particle velocity
V_{max}	5	Scalar, max. for particle velocity

The value of the fitness function with respect to number of iteration (*gbest*) is shown in Fig. 5.1. The final dimensions (in centimetre) of the spacer layers for reflection below –20 dB over a bandwidth of 2–10 GHz were found to be 0.2195, 1.1443, 1.1872, 0.6875, 0.6672, 2.5502, and 2.9510. The reflection over the entire range of 2–10 GHz with the optimized parameters is given in Fig. 5.2. It is seen that the maximum reflection over this range is -21.0663 dB.

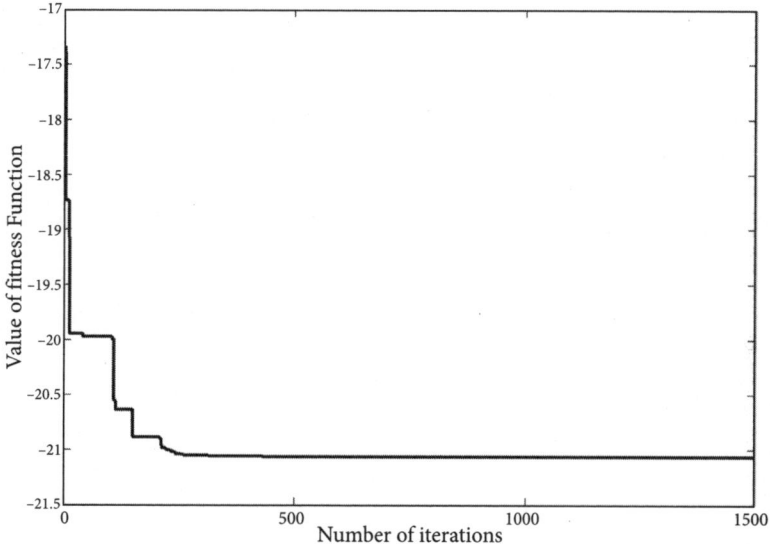

Fig 5.1 Variation of gbest value with respect to iterations for Jaumann RAM optimization

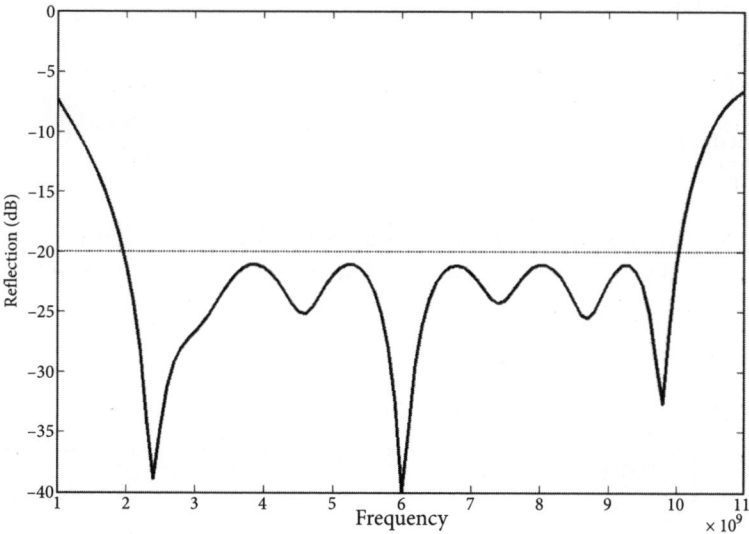

Fig 5.2 Reflection obtained in optimized Jaumann RAM design

However, at this juncture, it has to be mentioned that the algorithm can be extended to include optimization of selection of materials from a given database of materials and further optimization of thickness. This topic is discussed in the next section.

5.4.2 Multilayer RAM optimization

In the previous example, an optimized 14-layer Jaumann RAM was designed using particle swarm optimization. The optimization algorithm determined the thickness of each of the seven spacer layers. The material properties of each layer was fixed and known well in advance.

In this section, we aim to implement a more complex optimization algorithm; one that is capable of selecting materials from a database, arranging them and then determining the optimum thickness of each layer for best performance. Different types for materials like lossless dielectric materials, lossy dielectric materials, lossy magnetic materials, and relaxation type magnetic materials have been considered for the design.

Lossless dielectric materials are characterized by only real values of real part of permittivity (ε_r). The relative permeability of these materials is taken to be 1. A list of some typically used lossless dielectrics is given in Table 5.2.

Table 5.2 Lossless dielectric materials (DM)

No.	ε_r
1	10
2	50

When the dielectric material is lossy, there exists an imaginary component in the permittivity. These materials are often frequency dispersive. The real and imaginary parts of permittivity can be calculated as follows:

$$\varepsilon_r = \frac{\varepsilon_r(1GHz)}{f^\alpha} \tag{5.4}$$

$$\varepsilon_i = \frac{\varepsilon_i(1GHz)}{f^\beta} \tag{5.5}$$

where, f is the frequency of operation, and α and β are constants for a given material. Table 5.3 shows a list of typical lossy dielectric materials. The permittivity is then given by

$$\varepsilon = \varepsilon_r - j\varepsilon_i \tag{5.6}$$

Table 5.3 List of lossy dielectric materials (LDM)

	ε_r (1GHz)	α	ε_i (1GHz)	β
1	5	0.861	8	0.569
2	8	0.778	10	0.682
3	10	0.778	6	0.861

Lossy magnetic material can also be modelled using equations similar to those used for modelling lossy dielectric materials. The permittivity of these materials is considered to be constant and for the list of lossy magnetic materials given in Table 5.4, the permittivity is taken to be 15. The permeability of the lossy magnetic materials can be calculated using the following equations:

$$\mu_r = \frac{\mu_r(1GRz)}{f^a} \tag{5.7}$$

$$\mu_i = \frac{\mu_i(1GRz)}{f^\beta} \tag{5.8}$$

$$\mu = \mu_r - j\mu_i \tag{5.9}$$

where, f is the frequency of operation, and α and β are constants for a given material.

Table 5.4 List of lossy magnetic materials (LMM)

	μ_r (1GHz)	α	μ_i (1GHz)	β
1	5	0.974	10	0.961
2	3	1	15	0.957
3	7	1	12	1

Relaxation-type magnetic materials are also commonly used in RAM designs. These materials also possess constant permittivity. The permeability shows frequency dispersive property and can be modelled as follows:

$$\mu_r = \frac{\mu_m f_m^2}{f^2 + f_m^2} \tag{5.10}$$

$$\mu_i = \frac{\mu_m f_m f}{f^2 + f_m^2} \tag{5.11}$$

$$\mu = \mu_r - j\mu_i \tag{5.12}$$

The values for μ_m and f_m are generally specified with the material under consideration. Table 5.5 shows list of typical relaxation type magnetic materials. For all these materials, the permittivity is taken to be 15.

Table 5.5 List of relaxation-type magnetic materials (RLM)

	μ_m	f_m
1	35	0.8
2	35	0.5
3	30	1
4	18	0.5
5	20	1.5
6	30	2.5
7	30	2
8	25	3.5

The PSO algorithm selects 5 materials out of the materials given in the database and then varies their thicknesses in order to obtain minimum reflection. In order to do so, one calculates the reflection for either perpendicular polarization (TE) or parallel polarization (TM) of incident wave. The reflection for both these polarizations should be the same [Dib *et al.*, 2010]. Let the layers be numbered beginning from the innermost layer (the layer that touches the metallic structure the RAM is mounted on). Then, the generalized reflection coefficient at the interface of ith layer and the $(i-1)^{th}$ layer is given by

$$\tilde{R}_{i-1,i} = \frac{R_{i-1,i} + \tilde{R}_{i-2,i-1}\, e^{-2jk_i d_i}}{1 + R_{i-1,i} + \tilde{R}_{i-2,i-1}\, e^{-2jk_i d_i}} \tag{5.13}$$

where, d_i is the thickness of the ith layer and k_i is given as

$$k_i = \omega\sqrt{\mu_i \varepsilon_i - \mu_0 \varepsilon_0 \sin^2\theta} \tag{5.14}$$

where, ω is the angular frequency of the incident wave, μ_i and ε_i are the permeability and permittivity of the ith layer, respectively, μ_o and ε_o are the permeability and permittivity of free space, respectively, and θ is the angle of incidence of the wave [Dib *et al.*, 2010]. The value for $R_{i-1,i}$ is calculated using either of the following two expressions:

$$\text{TE: } R_{i-1,i} = \frac{\mu_{i-1}k_i - \mu_i k_{i-1}}{\mu_{i-1}k_i + \mu_i k_{i-1}} \tag{5.15}$$

$$\text{TM: } R_{i-1,i} = \frac{\varepsilon_{i-1}k_i - \varepsilon_i k_{i-1}}{\varepsilon_{i-1}k_i + \varepsilon_i k_{i-1}} \tag{5.16}$$

In the example, the reflection for TE mode is calculated. This corresponds to perpendicular polarization [Dib *et al.*, 2010].

For the PSO, the problem was defined to be a 10-dimensional one. The first five dimensions held values for material selection. Therefore, these dimensions were assumed to be integers ranging up to the total number of materials available in the database. The value of the integer represented the position of the material in the table. During the position update, the first fives terms in the dimensional matrix were allowed to be rounded off. The remaining five dimensions corresponded to the thickness of each layer. This value was allowed to range between [0.0001, 0.2] (in cm). Velocity update was performed using equation as given in Chapter 2. For increasing the performance of the algorithm, the social and cognitive parameters were assigned the value of 0.5 [Parsopoulos and Vrahatis, 2002] and the inertial weight was decremented from 0.9 at the start of the algorithm to 0.4 towards the end of the algorithm [Jin and Samii, 2007]. Optimization was performed for two different frequency ranges: 0.2–2 GHz and 0.1–10 GHz. The reflection for the entire range of frequencies was calculated using equations for each combination of material and thickness obtained in the algorithm. Then, the fitness was calculated using Eq. 5.3.

The algorithm was run for 300 particles and 300 iterations. For the first case, i.e., 0.2–2 GHz, the maximum reflection observed over the entire bandwidth was -31.255518 dB. The material number and the thickness of each layer are given in Table 1. Figure 5.3 shows the variation of the fitness function as the algorithm was run. Figure 5.4 shows the reflection obtained by designing a RAM using the parameters given in Table 5.6.

Table 5.6 Optimized material selection and thickness for 0.2–2 GHz

Layer	Material Number	Thickness (cm)
1	LMM-2	0.1835
2	LMM-2	0.2000
3	RLM-3	0.0334
4	LDM-2	0.1033
5	RLM-7	0.0927

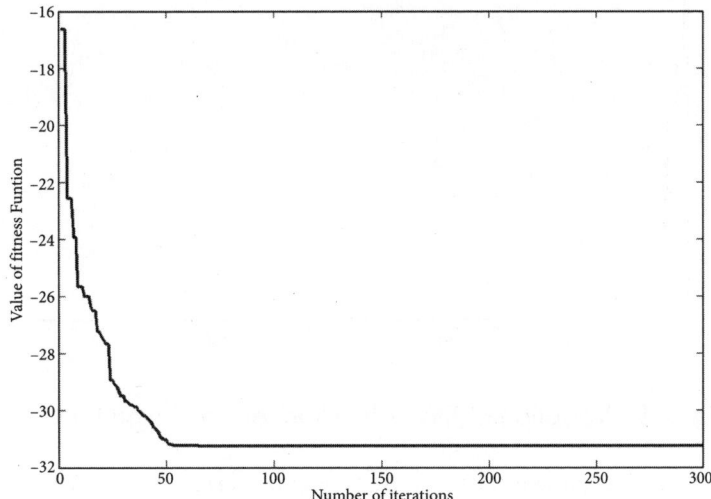

Fig 5.3 Variation of fitness function for simulation for 0.2–2 GHz

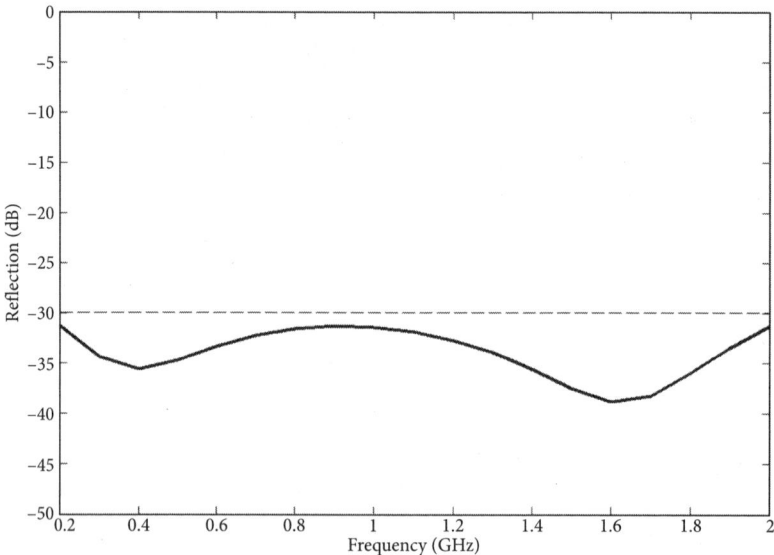

Fig 5.4   Reflection obtained for 0.2–2 GHz RAM design

The same algorithm was run for a frequency range of 0.1–10 GHz. The maximum reflection was found to be –18.885863 dB. The optimized design values are given in Table 5.7. Figures 5.5 and 5.6 provide the variation fitness function and reflection obtained using optimized values.

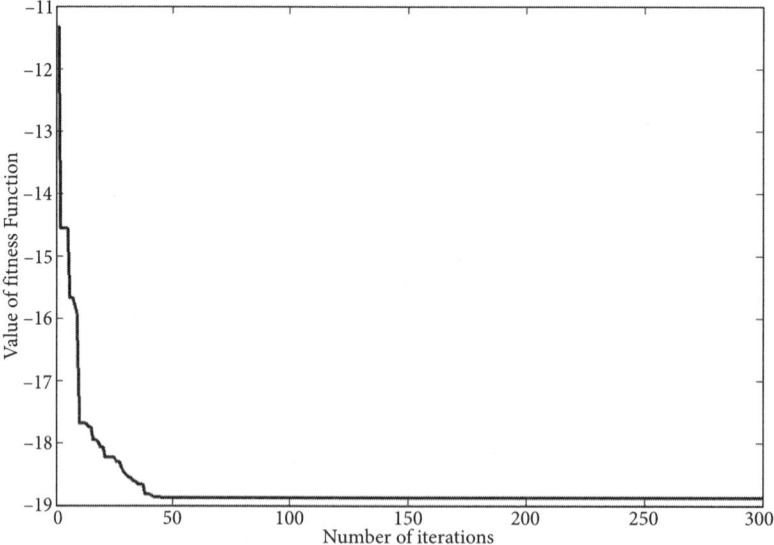

Fig 5.5   Variation of fitness function for simulation for 0.1–10 GHz

Material for RAM design cannot be chosen by considering the EM characteristics alone. Depending on the application, the designer must select materials with desired material properties such as strength, flexibility, etc. from a pre-determined database. The above mentioned method

Table 5.7 Optimized material selection and thickness for 0.1–10 GHz

Layer	Material Number	Thickness (cm)
1	RLM-4	0.0001
2	LMM-2	0.1838
3	LMM-3	0.1552
4	LDM-1	0.1546
5	RLM-6	0.0425

will simply the task of RAM engineering and is flexible enough to utilize a set of materials specified by the user.

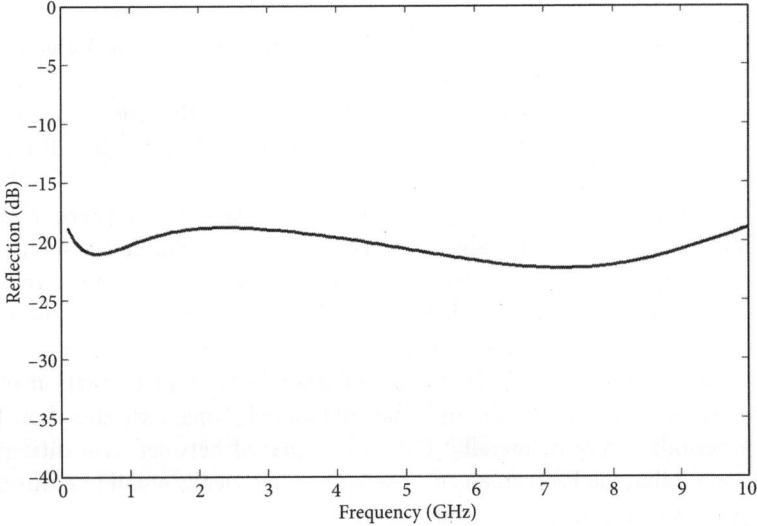

Fig 5.6   Reflection obtained for 0.1–10 GHz RAM design

5.5   Challenges and Issues in Conventional Absorber

The multilayer nature of conventional absorbers makes thickness one of the biggest design drawbacks. In narrowband absorbers, designs are based on quarter-wave transmissions lines for impedance transformation. A similar addition of thickness is observed in broadband absorbers as a large number of layers are required to realize broad-band characteristics. More material increases the weight of the absorber and this is highly undesirable in applications with strict mass budgets like aircraft. Further, some absorber designs require spacers with dielectric constant equal to that of air, for example, Styrofoam. These materials have poor structural properties and hence, cannot be used for rugged applications.

As a result, absorbers made up of metamaterials have gained popularity over the past few years. In this case, the metamaterial structure is designed in such manner that the permittivity and permeability become equal.

5.6   Microwave Metamaterial Absorber

The design of metamaterial absorbers is centred on fabricating the metamaterial structure in such a way that the relative permittivity is made equal to the permeability. This design procedure can be made more efficient by using soft computing techniques. In the coming section, the implementation of PSO for the design of metamaterial RAM is discussed.

5.6.1   Overview

New techniques for absorber design that are slightly different from the conventional design procedure, are gaining momentum [Kern and Werner, 2003a]. These techniques include the usage of electromagnetic band gap (EBG) structures, high impedance FSS surfaces, and artificial magnetic conductors (AMC). The design of these structures is a complex procedure, which can be simplified using soft computing.

A two-layer absorber whose structural parameters were optimized using genetic algorithm was designed by Wang et al. [Wang et al., 2008]. The design consisted of a left-handed material (LHM) backed by a high absorptivity material. It was observed that the inclusion of the LHM increased the bandwidth from 0.4 to 10 GHz. The maximum reflectance in this region was seen to be -12 dB.

A similar algorithm was used by Wang and Werner [Wang and Werner, 2009] for the design of a double-sided wideband absorber. The absorber contained resistive FSS screens and multi-band ultrathin EBG structures. The parameters of the structure were optimized for high absorption for different types of polarizations. The analysis was carried out using spectral domain periodic method of moments (PMM).

Bayraktar et al. [Bayraktar et al., 2010] demonstrated the implementation of the genetic algorithm for the optimization of a thin multi-layered metallo–dielectric absorber. The absorber consisted of a periodic array of metallic FSS screen placed between two different dielectric materials. It is noted that the FSS screen considered are electrically small in nature. The genetic algorithm targeted the optimization of the FSS layer.

The genetic algorithm was also used by Jiang et al. [Jiang et al., 2010b] to design a mid-infrared, metamaterial based absorber. The absorber consisted of three layers: the top-most layer was a patterned metallic layer, the middle layer consisted of a thin dielectric layer and the bottom layer was a metal sheet. The following cost function was used to vary the structural parameters of the absorber for absorptivity greater than 0.94

$$\text{Cost} = \sum\nolimits_{freq} \cdot \sum\nolimits_{\theta_i} \left[\left| A_{i,TE} - A_{tar} \right| + \left| A_{i,TM} - A_{tar} \right| \right] \tag{5.17}$$

where, $A = 1 - \left| S_{11} \right|^2 - \left| S_{21} \right|^2$ and the desired absorption i.e. $A_{tar} = 1$. $A_{i,TE}$ and $A_{i,TM}$ are the absorption for TE and TM polarization, respectively.

Another instance of the application of genetic algorithm was seen in the work reported by Kollatou et al. [Kollatou et al., 2011]. This group of researchers presented the design of an ultra-thin, improved bandwidth, wide angle metamaterial absorber with near perfect absorption. The GA optimized the structural parameters of the metamaterial. At 12.8 GHz, the design showed 99.2% absorption.

Micheli et al. [Micheli et al., 2011] designed a multi-layered, nano-structured absorber. GA was used to optimize this absorber. The GA code developed by these researchers was flexible enough to incorporate selection of algorithmic parameters like angle of incidence, frequency band, and overall thickness. This allows the user to attach priority to either minimization of thickness or maximization of losses.

Genetic algorithm can also be used to optimize electromagnetic smart screens designed to operate as absorbers as presented by Liu *et al.* [Liu *et al.*, 2011]. This smart screen was made up of a square patch loaded with pin diodes. The resultant structure behaved as a broadband absorber with a relative dynamic bandwidth of 40%. The same algorithm was used by Jiang *et al.* [Jiang *et al.*, 2011] to fabricate a conformal dual band metamaterial absorber. The absorber consisted of periodic stub-loaded H shaped nano-patches. The optimized design showed absorption greater than 90% at wavelengths 3.3μm and 3.9μm.

Guodos and Sahalos [Guodos and Sahalos, 2006] designed optimized multilayer planar absorptive coatings using multi-objective particle swarm optimization (MOPSO). The designed coating showed wide band and wide angle characteristics. Further, a comparison between MOPSO and multi-objective GA revealed that MOPSO was more efficient and required lesser computing time.

5.6.2 Design of microwave metamaterial absorber

A metamaterial absorber is a multilayer structure in which one of the layers is a metamaterial layer. As is well known, metamaterial structures are periodic EM designs consisting of metallic patterns printed on dielectric substrates. The literature survey reveals that a standard metamaterial RAM design provides upto 95% absorption. This performance can be improved further by optimizing the RAM design.

In this section, an attempt has been made to design an optimized metamaterial RAM where the structural parameters of the metamaterial layer have been extracted using soft computing techniques.

The well-known circular split ring resonator has been chosen as the candidate metamaterial unit cell towards design of the metamaterial based absorber. The schematic of a circular SRR with the dimensions is shown in Fig. 5.7a where r_{ext} is the external radius, w denotes the width

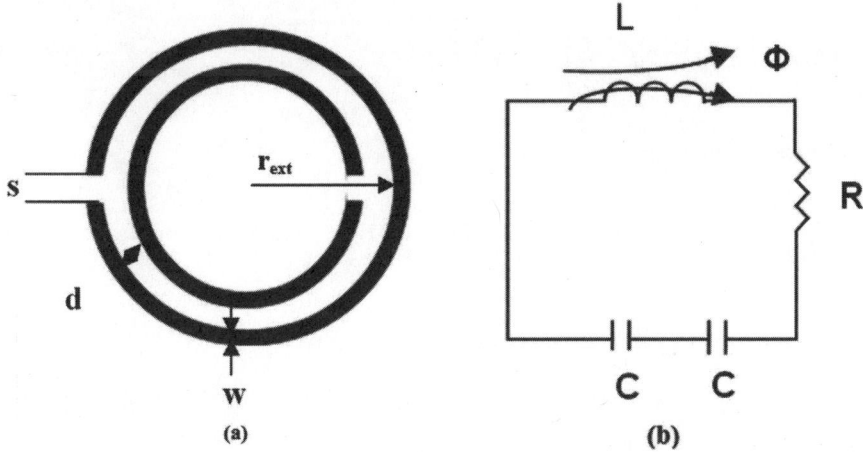

Fig 5.7 (a) Schematic diagram of circular SRR (b) Equivalent circuit of circular SRR

of rings, d is the gap present between the rings, and s represents the width of the split in the ring. The resonant frequency of the metamaterial structure depends on these structural parameters and therefore constitutes the basic requirement for efficient design. PSO in conjunction with equivalent circuit analysis (ECA) method as the EM tool has been used here to extract these structural parameters for the desired resonant frequency. In the ECA method, the distributed network is converted to lumped network (Fig. 5.7b) and analysis is carried out [Baena *et al.*, 2005].

The resonant frequency for circular SRR is given by

$$f_0 = \frac{1}{2\pi\sqrt{LC_S}} \tag{5.18}$$

If $r_{ext} < 5.2$ then the resonant frequency is [Pradeep *et al.*, 2011]

$$f_0 = \frac{1}{2\pi\sqrt{LC_S}} + \frac{5.2 - r_{ext}}{2} \tag{5.19}$$

If $r_{ext} > 5.2$ then the resonant frequency is

$$f_0 = \frac{1}{2\pi\sqrt{LC_S}} \tag{5.20}$$

where, L and C are the inductance and capacitance of the SRR, respectively. The expressions for L and C are given below.

The inductance L is given by

$$L = 2.57 e^{\frac{-w_3}{\sqrt{2}}} \left(\pi r_{ext} - 2.2d_1 - \frac{\pi}{2} \right) \tag{5.21}$$

The capacitance C is given by

$$C = 0.217 + \left\{ \frac{\left[0.059\left(2r_{ext} + \varepsilon_r - 5\right)\right]\left(0.437w_1 - 0.317w_2^2 + 0.07w_2^3\right)}{\left(3.3367e^{-3.2d_1} - 0.1955e^{-0.47h}\right)} \right\}$$
$$+ \left(0.05\varepsilon_r - 0.218\right) + \left(\frac{0.599h}{0.0248 + h} - 0.599 \right) \tag{5.22}$$

where, for $d < 1$ mm, $w_1 = w$, $w_2 = w$, $w_3 = w$ and $d_1 = d$.

The Matlab code for this ECA analysis has been developed which is the basic computational code for the PSO algorithm given in Chapter 2, Section 2.3.

5.6.3 PSO implementation

Proper selection of fitness function is crucial for efficient optimization. The fitness function selected for this application is given below.

$$f_{err} = \frac{|f_d - f_c|}{f_d} \qquad (5.23)$$

where, f_d is the desired frequency and f_c is the frequency obtained by using the equivalent circuit analysis method (Eqs. 5.18–5.22). The PSO algorithm incorporates this fitness function to extract the optimized structural parameters for the desired resonant frequency. PSO algorithm was run for 10 particles and 20 iterations. The inertial weight was taken to be a constant value of 0.25. The cognitive parameter, $c1$ and social parameter, $c2$ were taken to be 2.05 each. The maximum value of each dimension was set to 10 while the velocities were allowed to range from −1.5 to 1.5. The algorithm assumed the thickness of the substrate and its permittivity to be an *apriori* requirement and optimized the radius of the outer ring, width of each ring and distance between the rings.

5.6.4 Simulation results and discussion

The PSO algorithm was run using the defined parameters and with Eq. 5.10 as the fitness function. After optimization, for a resonant frequency of 1.67 GHz, the radius of the outer ring was found to be 5.2 cm, the width of the rings was found to be 0.3 cm, the distance between the two rings was found to be 0.4 cm, and the length of the splits was assigned a value of 1 cm. The substrate was made of polyimide and had a dimension of 5.3 cm × 5.3 cm × 0.85 cm. In order to design the absorber, an additional dielectric spacer of dimension 5.3 cm × 6 cm × 0.85 cm was added in the direction of wave propagation. This assembly was backed by a ground plane as shown in Fig. 5.8.

The absorption characteristics of this absorber were plotted by obtaining the reflection and then using Eq. 5.2. This is shown in Fig. 5.8b. At the resonant frequency, the absorption was found to be 99.64%.

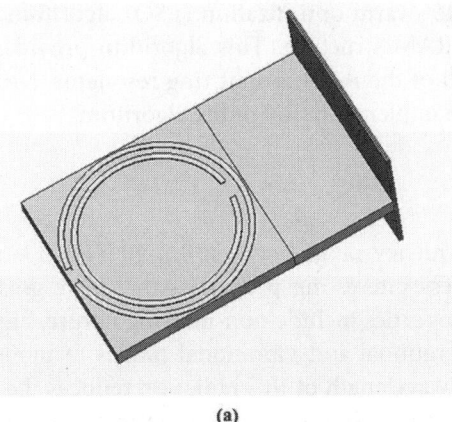

(a)

Fig 5.8(a)   Metamaterial absorber with circular SRR unit cell,

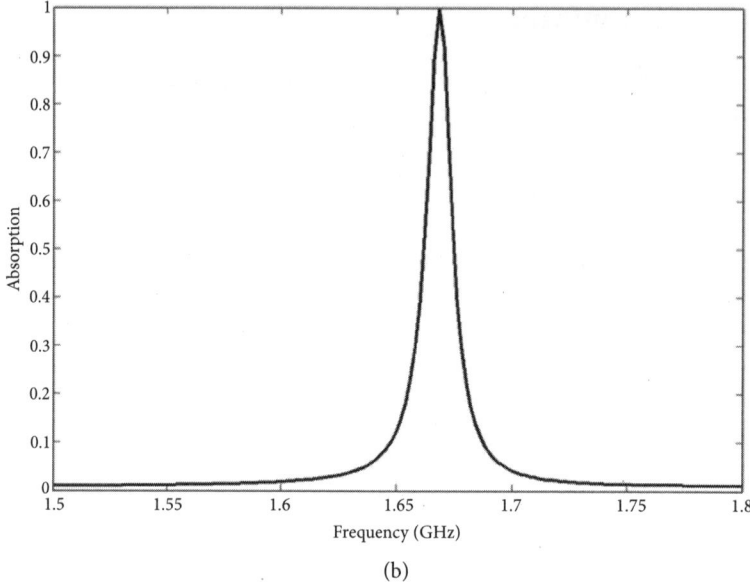

(b)

Fig 5.8(b)   Absorption characteristic of designed metamaterial absorber

5.7   Terahertz Absorber Design for Biomedical Application

The use of absorbers is not restricted to microwave regimes. With advances in terahertz imaging, the demand for terahertz absorbers is increasing. These absorbers find a place in the design of bolometer based detectors [Kearney, 2013] and studies have shown that this inclusion increases the efficiency of detectors. RAM for biomedical imaging requires wide bandwidth operation [Fitzgerald et al., 2006]. However, conventional RAM designs like Jaumann surfaces, Dallenbach, and Salisbury screens, etc., cannot be used as the terahertz region represents a region, which shows a natural breakpoint in material properties.

In this section, a particle swarm optimization (PSO) algorithm is applied to optimize the design of a metamaterial RAM structure. This algorithm provides the optimized structural parameters for the unit cell of the RAM, a split ring resonator. Near unity absorption at 1.16 THz was observed after the implementation of the algorithm.

5.7.1 Overview

Terahertz refers to the frequency range between 0.1–10 THz. Electromagnetic waves in this region are known to possess interesting properties that have great potential in the field of medical imaging. These properties include non-ionising nature, high degree of attenuation in water, ability to excite liberational and vibrational modes in molecules, etc. [Wallace et al., 2002]. The sub-millimetre wavelength of this radiation reduces the diffraction limit (given by 1.22 λ), thereby generating high resolution images of biological samples. This technology has already been used successfully to image and characterize human DNA [Fitzgerald et al., 2006].

However, while high resolution during imaging is a desirable property, the non-ionising property of terahertz radiation is perhaps the biggest reason for increased efforts towards making terahertz technology feasible commercially. Research has shown terahertz to be harmless to human tissues—low power exposure of biological samples did not divide keratinocytes [Wallace *et al.*, 2002]. Another interesting property that has found application in the biomedical field is the attenuation of terahertz radiation in water, the order of which lies between 100–1000 cm⁻¹ [Siegel, 2004]. While this property has severely restricted the usage of terahertz in long distance communication and radar applications, it can be used to develop high contrast images due to differential attenuation of radiation amongst various tissues. In fact, it has been observed that absorption of terahertz is higher in cancerous tissues than normal tissues [Fitzgerald *et al.*, 2006]. This has been attributed to the fact that cancerous cells have higher water levels than normal ones [Kearney, 2013].

Therefore, advances in terahertz spectroscopy have gained momentum with efforts being made to develop efficient systems for the same. The basic components of such systems include a terahertz source, a transmitter and a detector to measure the amount of radiation reflected off the tissue [Kearney, 2013]. The efficiency of the detector is, hence, key to the overall performance of the system. This requirement brings about the need for radar absorbing materials (RAM) that operate in the terahertz region.

Designing a RAM for terahertz is extremely challenging due to the lack of efficient EM responses in naturally occurring materials. This lack of availability is a consequence of a natural breakpoint of electric and magnetic properties of materials exposed to terahertz [Smith *et al.*, 2004]. Therefore, artificially engineered materials, called metamaterials, are used to overcome this difficulty. Metamaterial RAM designs that demonstrate efficiencies between 70–90% have been reported in literature [Wen *et al.*, 2009; Landy *et al.*, 2008; Landy *et al.*, 2009].

5.7.2 Biomedical spectroscopy system

Terahertz spectroscopy in the field of medicine is one of the most important applications of terahertz. The non-ionizing property of this radiation enables non-invasive study and characterization of human tissues, and the detection of various skin diseases.

Two types of terahertz instruments are available commercially:

- Passive terahertz instruments
- Active terahertz instruments

Passive terahertz instruments produce incoherent radiation. The images generated using this kind of radiation are not very accurate and hence is not used in bio-medical applications. However, some companies have already designed airport security systems that employ passive terahertz imaging. This would enable the detection of concealed items like knives.

On the other hand, active terahertz instruments generate coherent radiation either in the pulsed or continuous wave form. Two of the popular instruments used for active imaging include the free electron laser (FEL) and terahertz time domain spectroscopy (THz–TDS). FEL is capable of generating wide bandwidth, high power, and coherent radiation. This is done by passing an electron beam through sets of permanent magnets that are placed such that the H-fields of two adjacent pairs are opposite to each other. As a result, the direction of force acting on the electrons changes direction as they move from the influence of one pair of magnets to the other. The electrons oscillate and generate terahertz radiation. However, the instrument is complex, huge and expensive to construct. As a result most medical instruments employing terahertz technology use terahertz time domain spectroscopy (THz–TDS). Figure 5.9 shows a typical terahertz time domain spectroscopy system. A laser is used to generate extremely short pulses of time period approximately 90 fs. The pulse is split into two: one is sent to a delay line while the other is made to incident on a highly biased GaAs wafer. The latter configuration, called the Auston switch generates pulsed terahertz radiation. The signal from the delay line is later used as a probe pulse at the detector. The sample is then exposed to this pulsed terahertz signal. At the detector, the signal from the sample under test is sampled using the laser probe pulse from the delay line. This probe pulse samples the E-field of the terahertz signal and the change in the polarization of the probe pulse is directly proportional to the E-field.

From the block diagram, it is inferred that the performance of the detector is crucial to the overall imaging capability of the TDS system. Detectors have to be compact and inexpensive along with being highly efficient. Uncooled bolometers have found a comfortable niche in the domain of detectors due these properties, especially in thermal imaging [Kearny, 2013]. These detectors work on the principle of change in electrical resistance with temperature, i.e., the incident thermal energy. Indeed, micro-bolometer focal plane arrays are one of the most popular IR detectors available.

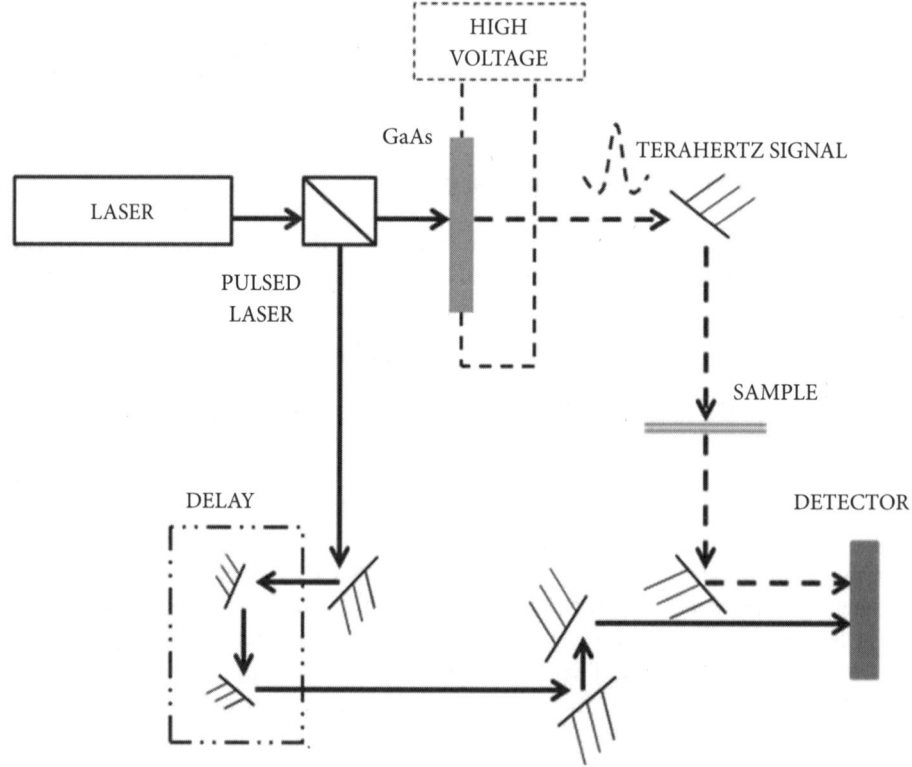

Fig 5.9 Block diagram of a typical terahertz spectroscopy system

Due to the proximity of the IR range of frequencies to terahertz in the electromagnetic spectrum, the same micro-bolometer detectors can be used in terahertz imaging albeit with a lower performance. This is due to the fact that the noise equivalent power (NEP) in the terahertz range is approximately 300 pW/Hz½ while that in IR is 14 pW/Hz½ [Kearny, 2013]. In other words, these detectors show poor responsivity in terahertz. Studies have shown that performance of these detectors can be improved significantly using absorbers.

Conventional absorber designs are based on multiple layers of resistive sheets and dielectric spacers. However, as discussed previously, naturally available materials do not show dielectric properties in the terahertz regime. Therefore, a need for using artificially engineered materials, called metamaterials, arises. The inclusion of these metamaterial based absorbers into bolometers will increase the sensitivity of these detectors in the terahertz region of the spectrum.

5.7.3 Design of metamaterial based terahertz absorber

A four layer RAM is designed here. A gold (conductivity of 4.09×10^7 S/m) split ring resonator (SRR) array is etched onto a dielectric spacer. Another sheet of gold is placed on the opposite face and is sandwiched in-between another spacer. The thickness of each spacer is taken to be 8 μm and the thickness of gold is taken as 0.4 μm. It should be noted here that the gap in the split ring contributes to the capacitance of the metamaterials while the linear bars behave as inductances. This LC combination effectively acts at a resonator and the resonant frequency depends on this LC combination. Hence, particle swarm optimization is used to vary the structural parameters, thereby tuning inductance and capacitance, and ultimately the resonant frequency.

In this discussion, a square split ring resonator (two square rings with a gap) is considered [Billoti, et al., 2007]. The SSRR is printed onto an 8 μm thick polyimide dielectric substrate with a permittivity of 2.8814. a represents the length of the side of the square, w is the width of the conductor, d is the space between the two squares, and g is the gap length (Fig. 5.10a). The equivalent LC representation is shown in Fig. 5.10b. Equivalent circuit analysis is carried out by converting the distributed elements into a lumped network and the resonant frequency is calculated to be

$$f_r = \frac{1}{2\pi\sqrt{LC_s}} \tag{5.24}$$

where, L is the effective inductance due to the rings and C_s is the equivalent capacitance, which can be calculated using Eqs. 5.25–(5.27 [Billoti et al., 2007].

$$L = \frac{4.86\mu_0}{2}(a-w-d)\left[\ln\left(\frac{0.98}{\rho}\right)+1.84\rho\right] \tag{5.25}$$

$$\rho = \frac{w+d}{a-w-d} \tag{5.26}$$

$$C_S = \left(a - \frac{3}{2}(w+d) \right) C_{pul}$$ (5.27)

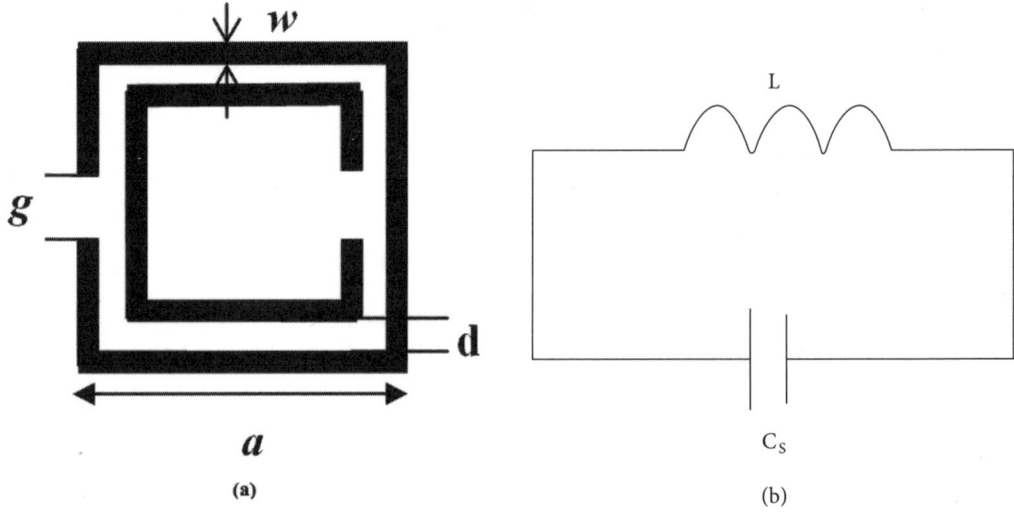

Fig 5.10 (a) Structure of square SRR, (b) LC Equivalent of Square SRR

Here, ρ is called the filling factor of inductance. C_{pul} is the per unit length capacitance between the rings.

5.7.4 Performance enhancement using PSO

The algorithm extracts the structural parameters of the square SSR by optimizing the fitness function given in Eq. 5.10. The algorithm was run for an *a priori* frequency of 1.16 THz. The extracted structural parameters are given in Table 5.8.

Table 5.8 Extracted geometry of square SRR

Parameter	Value (μm)
Side length, *a*	25.90
Width, *w*	3
Distance between rings, *d*	3
Gap length, *g*	0.3

Once the structural parameters are obtained, the RAM is simulated and the scattering parameters of the metamaterial RAM, S_{11} and S_{21}, are obtained. Using the scattering parameters and Eqs. 5.28– 5.31, the material parameters, viz. permittivity and permeability are extracted [Smith, *et al.*, 2002]. Let n denote the refractive index of the metamaterials RAM. It should be stressed here that this n is different from the n used in the algorithm. However, we wish to use the notation in order to conform to standard notations used in literature. Further, k denotes the angular wavenumber of the incident wave, z denotes the impedance of the metamaterial, and d_z denotes the thickness of the metamaterial in the direction of wave propagation.

$$n = \frac{1}{kd_z} \cos^{-1}\left[\frac{1}{2S_{21}}\left(1 - S_{11}^2 + S_{21}^2\right)\right] \tag{5.28}$$

$$z = \sqrt{\frac{\left(1 + S_{11}^2\right)^2 - S_{21}^2}{\left(1 - S_{11}^2\right)^2 - S_{21}^2}} \tag{5.29}$$

$$\varepsilon = \frac{n}{z} \tag{5.30}$$

$$\mu = nz \tag{5.31}$$

The simulated S parameters, and extracted permittivity and permeability of the structure are given in Fig. 5.11a and Fig. 5.11b, respectively.

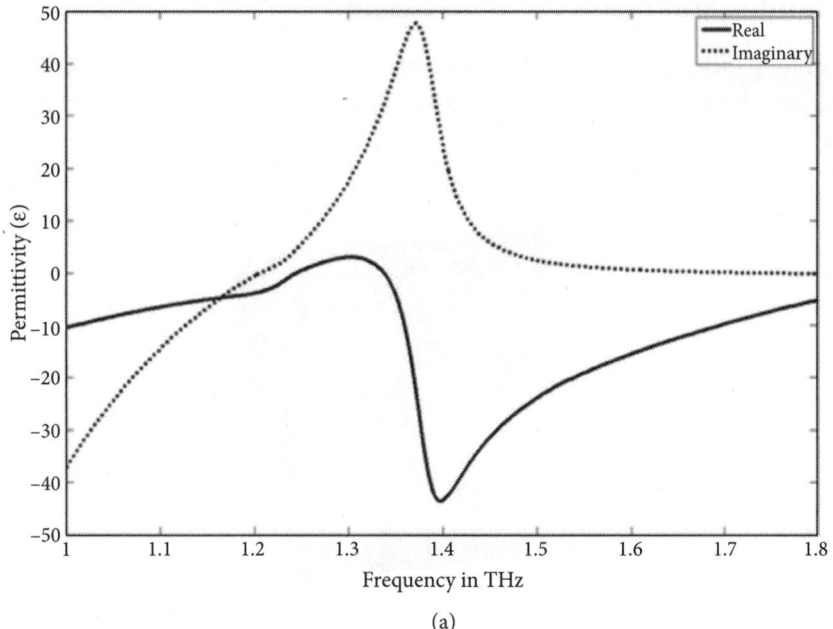

(a)

Fig 5.11(a)   Extracted permittivity of the designed square SRR

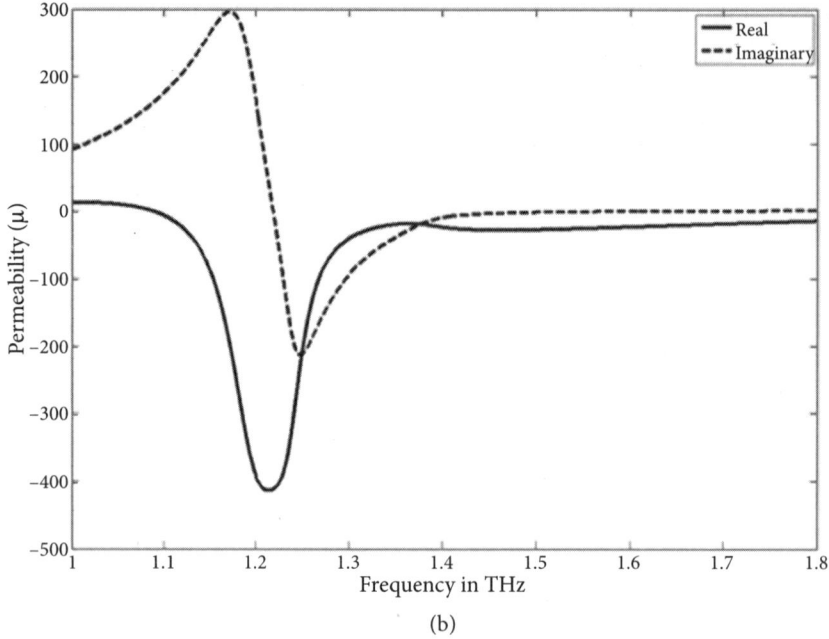

(b)

Fig 5.11(b)   Extracted permeability of the designed square SRR

5.7.5   Simulation results and discussion

An array of four such square split ring resonators was used to obtain the absorber (Fig. 5.12). The absorption characteristics for the same were simulated for different angles of incidence and the results were plotted (Fig. 5.12b).

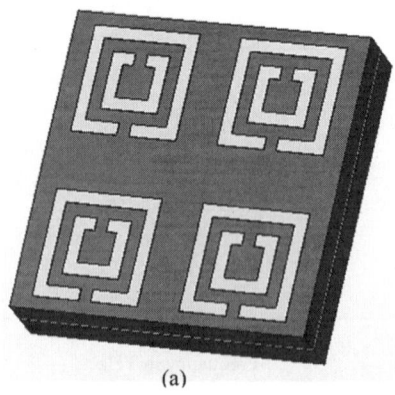

(a)

Fig 5.12(a)   The four-layer Metamaterial RAM design: The top layer consists of four optimized square SRR structures

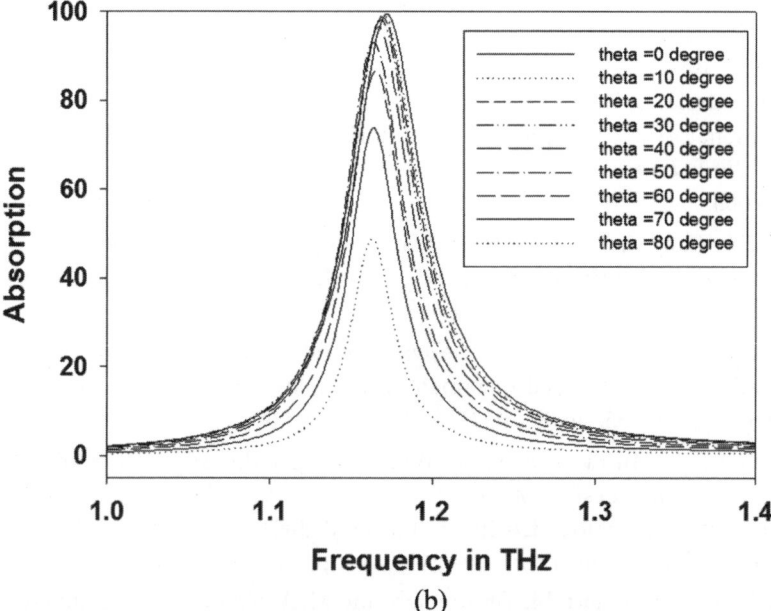

Fig 5.12(b) The four-layer Metamaterial RAM design: Absorption characteristics of PSO optimized RAM for different angles of incidence

From the simulation results, it is observed that the absorption at 1.16 THz was 99.32%. This performance establishes the effectiveness of particle swarm optimization algorithm for the generation of structural parameters of radar absorbing materials (RAM) for a desired frequency. The obtained bandwidth and high absorption make this designed RAM ideal for applications in terahertz biomedical imaging.

5.8 Summary

Design of multi-layer RAM is a complicated procedure and involves selection of a correct combination of materials followed by optimization of thickness in order to achieve best performance. In the chapter, particle swarm optimization has been established to be an effective tool for RAM design and optimization. Hybrid PSO can accommodate selection and placement of material, followed by optimization of thickness of each layer. This procedure can be used to design RAMs that show desirable characteristics over wide range of frequencies.

In addition to the conventional multi-layer RAM design, the chapter also probes the design of metamaterial based RAM for microwave as well as terahertz operation. PSO has been used to arrive at the optimized structural parameters of the metamaterial unit cell for a particular resonant frequency.

The RAM designs discussed in the chapter are planar in nature. For practical application in the aerospace field, RAMs often need to be mounted on curved surfaces. Therefore, the application of particle swarm optimization could be extended to include conformal RAM design optimization. Additionally, multi-objective optimization can be implemented in order to design tunable radar absorbers with embedded active elements for wide bandwidth.

References

Baena, J. D., J. Bonache, F. Martin, R. M. Silero, F. Falcone, T. Lopetagi, M. A. G. Laso, J. Garcia-Garcia, I. Gil, M. F. Portilo, and M. Sorolla, "Equivalent-circuit models for split ring resonators and complementary split ring resonators coupled to planar transmission lines," *IEEE Transactions on Microwave Theory and Techniques,* vol. 53, pp. 1451–1461, Apr. 2005.

Bayraktar Z., X. Wang, and D. H. Werner, "Thin composite matched impedance magneto-dielectric metamaterial absorbers," *Proceedings of IEEE Antennas and Propagation Society International Symposium,* pp. 1–4, Jul. 2010.

Bilotti, F., A. Toscano, and L. Vegni, "Design of spiral and multiple split-ring resonators for the realization of miniaturized metamaterial samples," *IEEE transactions on Antennas and Propagation,* vol. 55, no. 8, pp. 2258–2267, Aug. 2007.

Chakravarty, S., R. Mittra, and N. R. Williams, "On the application of the microgenetic algorithm to the design of broad-band microwave absorbers comprising frequency-selective surfaces embedded in multilayered dielectric media," *IEEE Transactions on Microwave Theory and Techniques,* vol. 49, pp. 1050–1059, Jun. 2001.

Chamaani, S., S. A. Mirtaheri, M. Teshnehlab, and M. A. Shooredeli, "Modified multi-objective particle swarm optimization for electromagnetic absorber design," *Proceedings of Asia Pacific Conference on Applied Electromagnetics,* 5p., Dec. 2007.

Choudhury, B., S. Bisoyi, and R. M. Jha, "Emerging trends in soft computing for metamaterial design and optimization," *Computers, Materials & Continua,* vol. 31, no. 3, pp. 201–228, 2012

Cui, S. and D. S. Weile, "Application of a parallel particle swarm optimization scheme to the design of electromagnetic absorbers," *IEEE Transactions on Antennas and Propagation,* vol. 53, pp. 3614–3624, Nov. 2005.

Dib, N., M. Asi, and A. Sabbah, "On the optimal design of multilayer microwave absorbers," *Progress In Electromagnetics Research C,* vol. 13, pp. 171–185, 2010.

Fitzgerald, A. J. F., V. P. Wallace, M. Jimenez-Linan, L. Bobrow, R. J. Pye, A. D. Purushotham, and D. D. Arnone, "Terahertz pulsed imaging of human breast tumors," *Radiology,* vol. 239, no. 2, pp. 533–540, May 2006.

Goudos, S. K. and J. N. Sahalos, "Microwave absorber optimal design using multi-objective particle swarm optimization," *Microwave and Optical technology letters,* vol. 48, no. 8, pp. 1553–1558, Aug. 2006.

Jiang, Z. H., Q. Wu, X. Wang, and D. H. Werner, "Flexible wide-angle polarization-insensitive mid-infrared metamaterial absorbers," *Proceedings of IEEE Antennas and Propagation Society International Symposium,* pp. 1–4, Jul. 2010.

Jiang, Z. H., S. Yun, F. Toor, D. H. Werner, and T. S. Mayer, "Experimental demonstration of a conformal optical metamaterial absorber," *Proceedings of IEEE Antennas and Propagation Society International Symposium,* pp. 1812–1815, 2011.

Jin, N. and Y. R. Samii, "Advances in particle swarm optimization for antenna designs: real number, binary, single objective and multiobjective implementations," *IEEE Transactions on Antennas and Propagation,* vol. 55, no. 3, pp. 556–567, Mar. 2007.

Kearney, B. T., *Enhancing microbolometer performance at terahertz frequencies with metamaterial absorbers*, Doctorate of Philosophy dissertation, 69 p., Naval Postgraduate School, 2013.

Kennedy, J., and R. Eberhart, "Particle swarm optimization," *Proceedings of IEEE International Conference on Neural Networks*, pp. 1942–1948, 1995.

Kern, D. J., and D. H. Werner, "A genetic algorithm approach to the design of ultra-thin electromagnetic bandgap absorbers," *Microwave Optical Technology Letters,* vol. 38, pp. 61–64, Jul. 2003.

Kollatou, T. M., A. I. Dimitriadis, N. V. Kantartzis, and C. S. Antonopoulos, "A bandwidth-enhanced, ultra-thin, wide-angle metamaterial absorber for EMC applications," *Proceedings of the 10th International Symposium on Electromagnetic Compatibility,* pp. 686–689, Sep. 2011.

Landy, N. I., C. M. Bingham, T. Tyler, N. Jokerst, D. R. Smith, and W. J. Padilla, "Design, theory, and measurement of a polarization insensitive absorber for terahertz imaging," *Physical Review B,* vol. 79, pp. 125104(1)–125104(6), 2009.

Landy, N. I., S. Sajuyigbe, J. J. Mock, D. R. Smith, and W. J. Padilla, "Perfect metamaterial absorber," *Physical Review Letters,* vol. 100, pp. 207402(1)–207402(4), May 2008.

Liu, H., L. Zhang, Y. Gao, Y. Shen, and D. Shi, "Electromagnetic wave absorber optimal design based on improved particle swarm optimization," *Proceedings of EMC'09,* pp. 797–800, Dec. 2009.

Liu, L., S. Matitsine, R. F. Huang, and C. B. Tang, "Electromagnetic smart screen with extended absorption band at microwave frequency," *Metamaterials 5,* pp. 36–41, 2011.

Liang, T., L. Li, J. A. Bossard, D. H. Werner, and T. S. Mayer, "Reconfigurable ultra-thin EBG absorbers using conducting polymers," *Proceedings of IEEE Antennas and Propagation International Symposium,* vol. 2B, pp. 204–207, Jul. 2005.

Michielssen, E., J. M. Sajer, S. Ranjithan, and R. Mittra, "Design of lightweight, broad-band microwave absorbers using genetic algorithms ," *IEEE Transactions on Microwave Theory and Techniques,* vol. 41, pp. 1024–1031, Jun. 1993.

Micheli, D., R. Pastore, C. Apollo, M. Marchetti, G. Gradoni, V. M. Primiani, and F. Moglie, "Broadband electromagnetic absorbers using carbon nanostructure-based composites," *IEEE Transactions on Microwave Theory and Techniques,* vol. 59, no. 10, pp. 2633–2646, Oct. 2011.

Parsopoulos, K. E. and M. N. Vrahatis, "Recent approaches to global optimization problems through Particle Swarm Optimization," *Natural Computing,* vol. 1, pp. 235–306, 2002.

Pradeep, A., S. Mridula, and P. Mohanan, "Design of an edge-coupled dual-ring split ring resonator," *IEEE Antennas and Propagation Magazine,* vol. 53, no. 4, pp. 45–54, Aug. 2011.

Robinson, J., Y. R. Samii, "Particle swarm optimization in electromagnetics," *IEEE Transactions on Antennas and Propagation,* vol. 52, no. 2, pp. 397–407, 2004.

Siegel, P. H. "Terahertz technology in biology and medicine," *IEEE Transactions on Microwave Theory and Techniques,* vol. 52, no. 2, pp., 2438–2447, Oct. 2004.

Smith, D. R., J. B. Pendry, and M. C. K. Wiltshire, "Metamaterials and negative refractive index," *Science,* vol. 305, pp. 788–792, Aug. 2004.

Smith, D. R., S. Schultz, P. Markoscanon, and C. M. Soukoulis, "Determination of effective permittivity and permeability of metamaterials from reflection and transmission coefficients," *Physical Review B,* vol. 65, pp. 195104 (1)–195104 (5), Apr. 2002.

Vinoy, K. J., and R. M. Jha, *Radar Absorbing Materials from Theory to Design and Characterization.* Kluwer Academic Publishers, Boston, ISBN 0-7923-9753-3, 1996.

Wallace, V. P., D. A. Arnone , R. M. Woodward, and R. J. Pye, "Biomedical applications of terahertz pulse imaging," *Proceedings of the Second Joint EMBS/BMES Conference,* pp. 2333–2334, Oct. 2002.

Wang, X., and D. H. Werner, "Multiband ultra-thin electromagnetic band-gap and double-sided wideband absorbers based on resistive frequency selective surfaces," *Proceedings of IEEE Antennas and Propagation Society International Symposium, APSURSI '09,* pp. 1–4, Jun. 2009.

Wang, Z., Z. Zhang, S. Qin, L. Wang, and X. Wang, "Theoretical study on wave-absorption properties of a structure with left and right handed materials," *Materials and Design,* vol. 29, no. 9, pp. 1777–17779, Oct. 2008.

Weile, D. S., E. Michielssen, and D. E. Goldberg, "Genetic algorithm design of pareto optimal broadband microwave absorbers," *IEEE Transactions on Electromagnetic Compatibility,* vol. 38, pp. 518–524, Aug. 1996.

Wen, Q. Y., H. W. Zhang, Y. S. Xie, Q. H. Yang, and Y. L. Liu, "Dual band terahertz metamaterial absorber: Design, fabrication, and characterization," *Applied Physics Letters*, vol. 95, pp. 241111(1)–241111(1), 2009.

Characterization of Planar Transmission Lines Using ANN

6

Advancements in *microwave integrated circuits* (MIC) are occurring at a rapid pace. Features of MIC such as small size and the possibility of realization of entire systems on a single chip have led to the increase in the usage of these devices in communication systems. One of the integral elements of these MICs is transmission lines. Transmission lines are not just used to connect individual modules inside the MIC but they are also essential in the realization of components like filters, mixers, couplers, and power dividers. Popular types of transmission lines used include microstrip lines, strip lines, grounded waveguides, slotlines, etc. As a result, these transmission lines must be designed carefully and analyzed; deviations in the design will lead to errors in the operation of the MIC. Traditional methods of analysis of transmission lines include conformal transformation, variational method, spectral domain method, and numerical methods like finite element method (FEM), finite difference method, and finite difference time domain method. These methods are extremely accurate and have been documented to produce results that concur with experimental observations. However, complex mathematical formulation and iterative nature imply high computational time requirement [Bhatt and Koul, 1990; Schellenberg, 1995]. This quality limits the application of these techniques in situations where quick solutions are required such as CAD packages for transmission line design. The algorithms based on artificial intelligence could be used in order to create such CAD packages. Yildiz *et al.* [Yildiz *et al.*, 2004] used neural network for analysis of co-planar waveguide. The same algorithm was further used for the analysis of inverted microstrip lines [Yildiz and Saracogulu, 2003]. Although the analysis of planar transmission lines using neural network has been explored, the design of various configurations of transmission lines using NN is yet to be done.

In this chapter, the analysis and the design of various configurations of transmission lines, namely the microstrip line and the slotline, are conducted using artificial neural networks. The corresponding formulation and ANN implementation are discussed in detail.

6.1   Planar Transmission Lines

As mentioned earlier, different types of transmission lines are used in the fabrication of MIC depending upon the requirement. In this chapter, the focus is limited to two types of planar transmission lines viz. microstrip transmission lines and slotline transmission line.

6.1.1 Microstrip lines

The most popular type of transmission line, the microstrip line, consists of a signal carrying metallic strip on a grounded dielectric substrate as shown in Fig. 6.1. These microstrip transmission lines can be fabricated using conventional PCB fabrication technology, and hence are cheap to manufacture. The expression for the impedance for a microstrip transmission line is given by [Wheeler, 1964]:

$$
Z = \frac{Z_o}{2\pi\sqrt{2(1+\varepsilon_r)}} \ln\left(1 + \frac{4h}{w_{eff}}\left(\frac{14+\frac{8}{\varepsilon_r}}{11}\frac{4h}{w_{eff}} + \sqrt{\left(\frac{14+\frac{8}{\varepsilon_r}}{11}\frac{4h}{w_{eff}}\right)^2 + \pi^2\frac{1+\frac{1}{\varepsilon_r}}{2}}\right)\right) \tag{6.1}
$$

where, Z_o is the free space impedance, w_{eff} is the effective width of the strip, which takes into account the actual width of the metallic strip (w) and a correction factor for finite thickness of the metal, h is the substrate thickness, and ε_r is the relative permittivity of the substrate. The value for w_{eff} can be calculated using

$$
w_{eff} = w + t\frac{1+\frac{1}{\varepsilon_r}}{2\pi}\ln\left(\frac{4e}{\sqrt{\left(\frac{t}{h}\right)^2 + \left(\frac{1}{\pi}\frac{1}{\frac{w}{t}+\frac{11}{10}}\right)^2}}\right) \tag{6.2}
$$

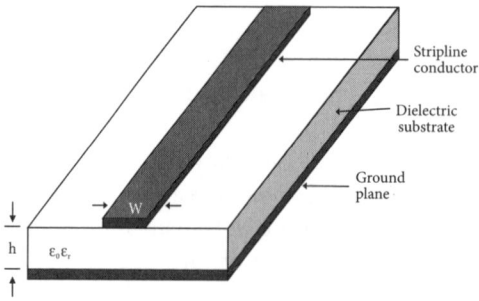

Fig 6.1   Schematic diagram of microstrip transmission line

6.1.2 Slot line transmission lines

As the name suggests, a slotline can be realized by creating a slot in a planar metallic strip as shown in Fig. 6.2. The presence of the slot implies that two metallic strips are now separated from each other. The two strips are responsible for the propagation of information through the transmission line. On comparing Figs. 6.1 and 6.2, it is evident that there is no separate ground plane in the realization of a slotline. As a result, slotlines are one sided, and this property is one of its biggest advantages. However, unlike the microstrip line, the power handling capability and unloaded quality factor, Q of a slotline is very low. Further, the radiation losses in a slotline have been observed to be higher than that seen in microstrip lines.

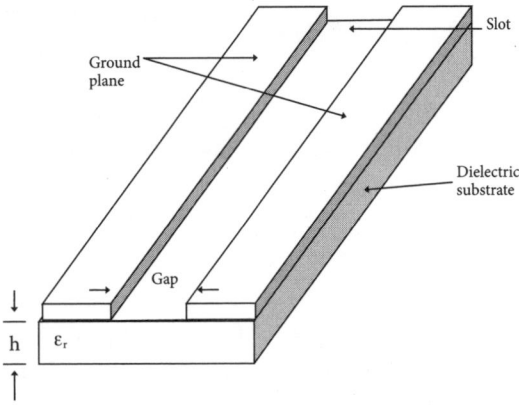

Fig 6.2 Schematic diagram of slotline transmission line

6.2   ANN Implementation

Artificial neural networks have been used for modelling, simulation, and optimization of transmission lines. The basis for the development of these networks has been measured and/or simulated data from various experiments. This data helped create accurate, quick, and reliable neural networks [Devabhaktuni et al., 2001]. In this section, a neural network algorithm has been developed for the design and analysis of microstrip and slotlines transmission lines [Pattanayak et al., 2009].

The theory behind neural networks has been discussed in Chapter 2. However, we will provide a brief overview of the same here. The development of a neural network involves three steps: generation of data, training of the neural network, and testing [Zhang and Gupta, 2000]. Initially, it is assumed that the neural network has no knowledge about the transmission lines and its design. Therefore, it must be trained using data generated from reliable sources. The trained neural network is then tested for the desired accuracy. If the errors are minimal, the neural network can be used without the training network. The flowchart for development of the neural network model is given in Fig. 6.3.

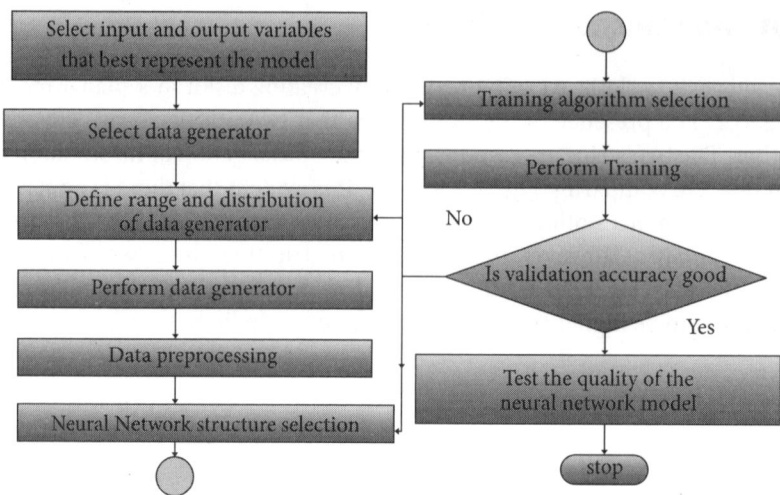

Fig 6.3   Flowchart of the neural network model for design and analysis of transmission lines

6.2.1 Generation of data

Since the neural network is trained using data determined from other sources, the accuracy of this data is crucial to the reliable performance of the neural network. For a particular set of input, the output was determined using a transmission line calculator.

6.2.2 Training of the neural network

The algorithm used for training is the backprogation algorithm described in Chapter 2. For a given set inputs, this algorithm tries to minimize the difference between the desired output and the output generated by the network. This is represented as:

$$D = \frac{1}{2}\sum_{j}(d_j - n_j)^2 \tag{6.3}$$

where, d_j is the desired output j and n_j is the actual output.

The back propagation algorithm feeds the error back into the network. The weights of the network are modified in order to reduce this error. Each weight w_{ji} is modified by adding Δw_{ji} to it so as to reduce D rapidly.

6.2.3 Testing

Once the neural network is trained and the optimized weights are determined, the network was tested for different input sets. When the outputs show the desired accuracy, the corresponding weights and biases are stored. The neural network can then be used without the training network.

6.3   Analysis and Design of Microstrip Transmission Line

This section focuses on the implementation of ANN for two different microstrip line based problem statements. These two problem statements include (i) analysis of microstrip line, and (ii) design of microstrip line. The analysis of microstrip transmission line includes determining the effective dielectric constant and the characteristic impedance with known physical dimensions. On the other hand, the design aspect involves determining the physical dimensions such as width and length of the transmission line for particular characteristic impedance.

6.3.1 Analysis of microstrip line

Wave propagation in a microstrip transmission line depends on its characteristic impedance and effective dielectric constant. These values are dependent on the physical parameters of the line itself. Determination of the characteristic impedance and effective dielectric constant is an effective way to analyze a transmission line. The neural network was developed to take height of substrate, permittivity of substrate, width and thickness of microstrip, conductivity of the material used for the fabrication of the microstrip, and frequency of operation as the inputs and provides the characteristic impedance and effective dielectric constant as output.

The analysis is carried out for a frequency range 0.1–1.5 GHz and for a substrate thickness in the range 0.5–2 mm. Different substrates like GaAs, alumina, silicon, RT/Duroid5880, and beryllium oxide, and conducting material like nickel, aluminium, copper, gold, and silver were also considered for the analysis. Table 6.1 gives the optimized neural network structure for this analysis and Table 6.2 gives the training parameters of the neural network. Figure 6.4 shows the variation in the accuracy of the neural network with time.

Table 6.1   Neural network structure for the analysis of microstrip transmission line

Layer	Number of Neurons	Sigmoid function
Input layer	6	Purelin
1st hidden layer	70	Tansig
2nd hidden layer	40	Tansig
3rd hidden layer	40	Tansig
Output layer	2	Purelin

Table 6.2   Training parameters considered for the neural network model for analysis of microstrip transmission line

Training Parameters	Values
Epochs	5000
Learning rate	0.01
Total data set	1800
Goal	0.001
Training algorithm	Resilent backpropagation

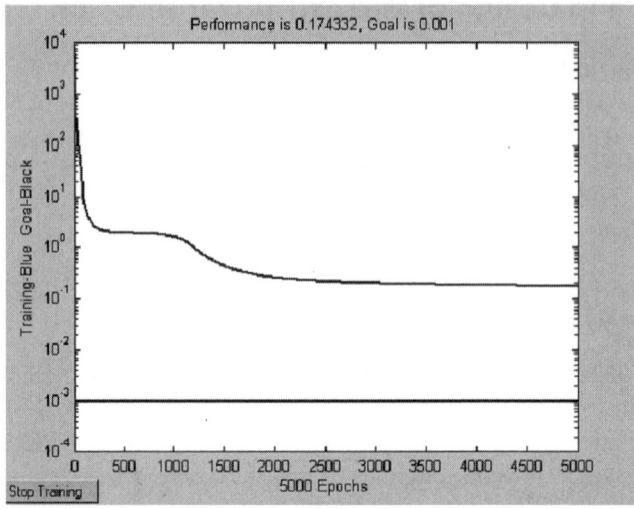

Fig 6.4 Training curve for analysis of microstrip transmission line

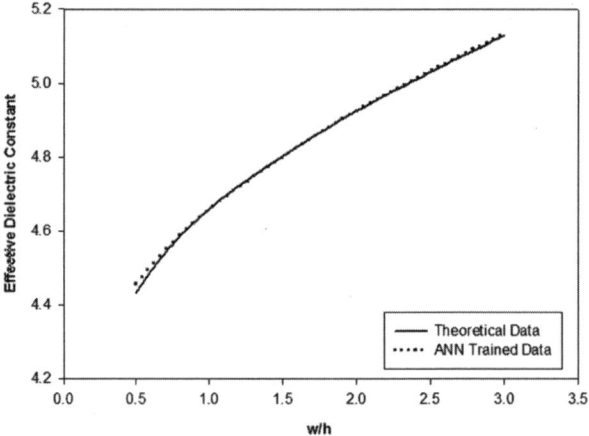

Fig 6.5 Effective dielectric constant vs width/height ratio of microstrip transmission line for a substrate of dielectric constant 6.7 (Theoretical model: Data generated by spectral domain method and neural network data)

The desired mean square error was set to 0.001. The neural network was able to realize a mean square error of 0.174332 (Fig. 6.4). As a test case, the variation in the effective dielectric constant with respect to the w/h (ratio between width and height) was plotted using the neural network. The generated data was compared with theoretical values obtained using spectral domain method. Figure 6.5 shows the comparison and it can be concluded that the data obtained from the two methods are in excellent agreement with each other.

6.3.2 Design of microstrip line

Globally, a standard impedance of 50 Ω is used in the design of high frequency systems. Almost of all input and output ports of high frequency systems are matched to this value of impedance. Consequently, most of the transmission lines are also designed for this value of impedance.

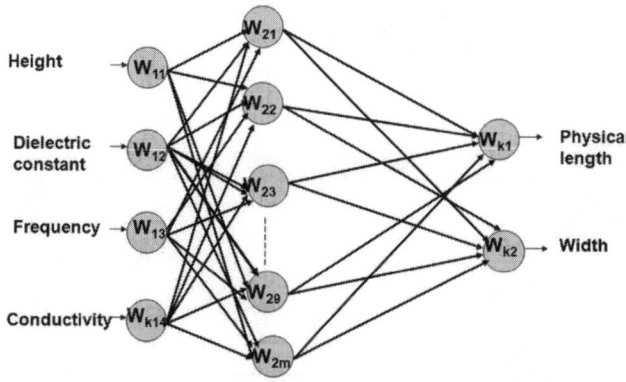

Fig 6.6 Schematic diagram of the neural network model for design of microstrip line

The impedance of a microstrip transmission line depends on its width and conductivity as well as the height and permittivity of the substrate. The length of the microstrip line does not affect its characteristic impedance. However, length becomes an important parameter for the realization of a particular phase. The schematic diagram of the neural network for design of microstrip line is shown in Fig. 6.6. The input set consists of height and dielectric constant of the substrate, frequency of operation, and conductivity of the metal used. The output of the neural network includes the length and width of the microstrip line.

The conditions for the development of the neural network are the same as the ones given in Section 6.4.1, viz., a frequency range from 0.1–1.5 GHz was chosen and substrate thickness was allowed to range from 0.5–2 mm. Different substrates like GaAs, alumina, silicon, RT/Duroid5880, alumina, and beryllium oxide, and conducting material like nickel, aluminium, copper, gold, and silver were also considered for the analysis. Tables 6.3 and 6.4 give the optimized neural network design for this problem statement and the training parameters respectively.

Table 6.3 Neural network for the design of microstrip transmission line

Layer	Number of Neurons	Sigmoid function
Input layer	4	Purelin
1st hidden layer	30	Tansig
2nd hidden layer	9	Tansig
Output layer	2	Purelin

Table 6.4   Neural network training for design of microstrip transmission line

Training Parameters	Values
Epochs	5000
Learning rate	0.1
Total data set	1800
Goal	0.001
Training algorithm	Resilent backpropagation

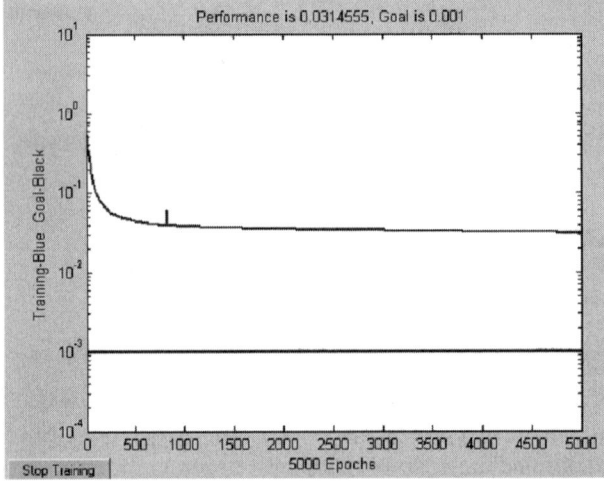

Fig 6.7   Training curve for design of microstrip transmission line

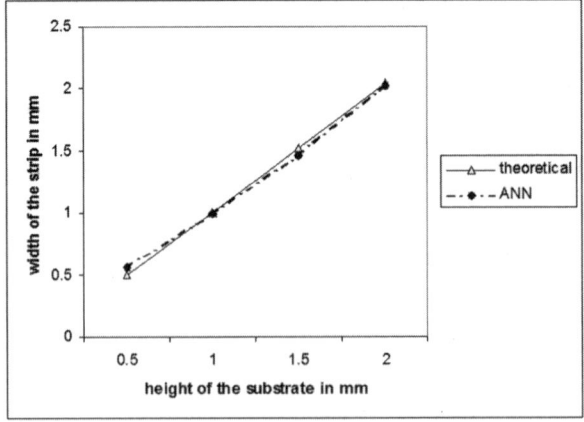

Fig 6.8   Comparison of theoretical data and neural network data for design of 50 Ω microstrip transmission line

Similar to the previous example, the desired accuracy for the neural network was set to 0.001. After training, a performance of 0.00314555 was realized. The variation in the performance of the neural network with respect to time is given in Fig. 6.7. The corresponding performance evaluation of the neural model has been carried out and the comparison between theoretical data and ANN data is given in the Fig. 6.8.

6.4   Analysis and Design of Slotline

The slotline is another popular type of transmission line that is often used instead of microstrips during the design of MICs. As described in the Section 6.4, the analysis of slot line transmission line includes calculation of the characteristic impedance and the effective dielectric constant, with the gap between the two ground plates, frequency of operation, dielectric constant of the substrate, and height of the substrate as *a priori* information. Similarly, design of the slotline corresponds to the determination of the gap and the physical length of the slot line. Hence, the procedures for analysis and design performed in Section 6.4 were repeated for a slotline transmission line.

6.4.1 Analysis of slotline

The frequency of operation, gap between the two conductors, height and dielectric constant of the substrate were considered to be the inputs to the neural network. The aim of the network was to arrive at the characteristic impedance and effective dielectric constant of the slotline.

Table 6.5 Neural network structure for the analysis of slotline

Layer	Number of Neurons	Sigmoid function
Input layer	3	Purelin
1st hidden layer	70	Tansig
2nd hidden layer	30	Tansig
Output layer	2	Purelin

Table 6.6 Training parameters of the neural network model for the analysis of slotline

Training Parameters	Values
Epochs	5000
Learning rate	0.001
Total data set	251
Goal	0.001
Training algorithm	Trainrp

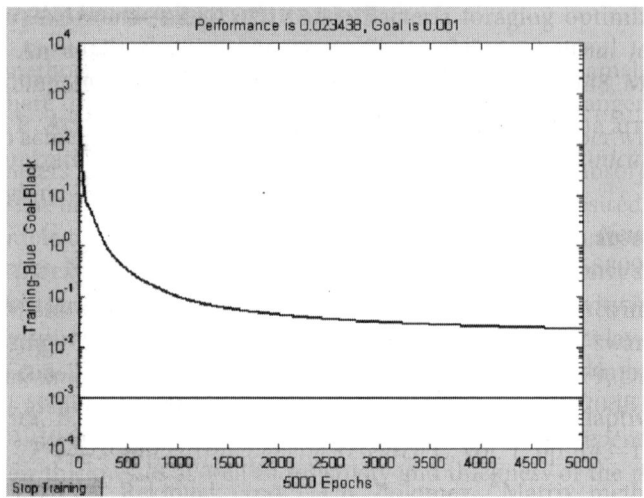

Fig 6.9   Training curve of analysis of slotline

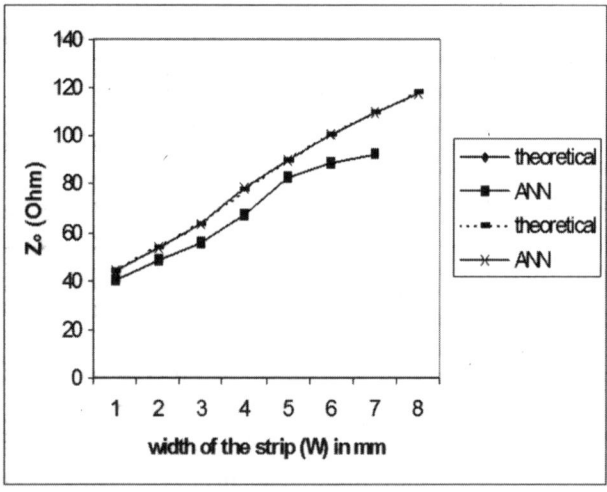

Fig 6.10   Characteristic impedance vs width of the slotline

The output from the neural network was generated for a frequency sweep of 0.5–2 GHz, slot range between 0.2–6 mm and height of substrate lying between 6–9 mm. The data was generated for different substrates like GaAs, silicon, RT/Duroid5880, and alumina. Table 6.5 gives the neural network structure obtained for this problem statement and Table 6.6 gives the corresponding training parameters. Just like the previous cases, the design goal was an error of 0.001 and the designed neural network was able to achieve a regularized mean square error of 0.023438. The variation in the performance of the neural network with respect to time is shown in Fig. 6.9. A comparative study of characteristic impedance vs width of the slot line is given in Fig. 6.10.

6.4.2 Design of slotline

The neural network is used to determine the dimensions of the gap and the length of the slotline. The inputs to the network are operating frequency, height and relative permittivity of the substrate. A frequency sweep from 0.5 and 2.5 GHz was conducted for a substrate height ranging from 6 and 8 mm. The final neural network structure for the problem statement is given in Table 6.7 and Table 6.8 shows the training parameters for the same. The neural network was seen to have a performance of 0.0099863 as seen in Fig. 6.11.

Table 6.7 Neural network structure for the design of slotline

Layer	Number of Neurons	Sigmoid function
Input layer	3	Purelin
1st hidden layer	70	Tansig
2nd hidden layer	30	Tansig
Output layer	2	Purelin

Table 6.8 Training parameters of the neural network model for the design of slotline

Training Parameters	Values
Epochs	5000
Learning rate	0.001
Total data set	251
Goal	0.001
Training algorithm	Trainrp

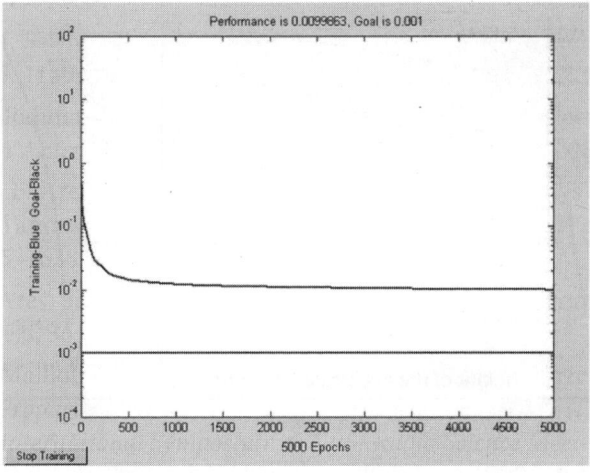

Fig 6.11 Training curve of the design of slotline

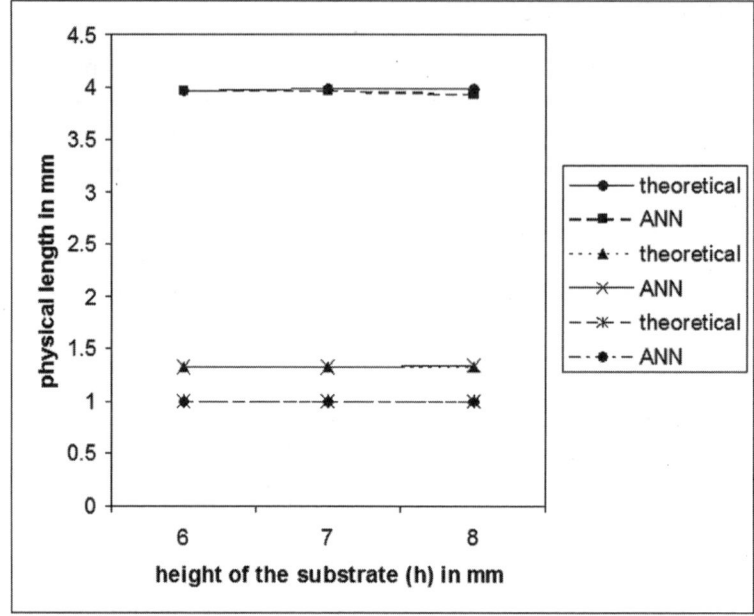

Fig 6.12   Design of slotline: Physical length vs height of the substrate

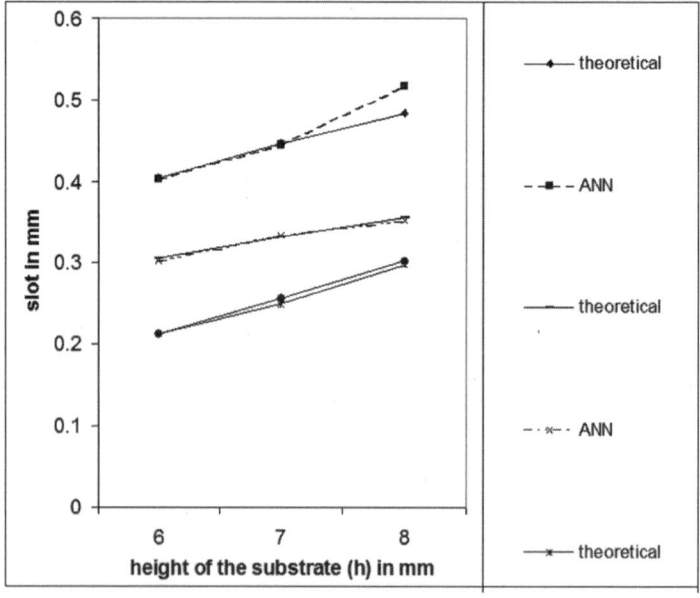

Fig 6.13   Design of slotline: Width of the slotline vs height of the substrate

6.5   Summary

Soft computing techniques provide robust, quick, and reliable solutions to otherwise computationally intensive problems. In the chapter, one such soft computing technique, viz. artificial neural network has been implemented, in order to analyse and design transmission lines. The focus was on two of the popular types of transmission lines viz. microstrip lines and slotlines. The design and analysis of these transmission lines through conventional methods is mathematically complex and hence extremely time consuming. Neural networks provide an option to bypass this complexity and arrive at accurate solutions in a time efficient manner.

References

Bhatt B. and S. K. Koul, *Stripline-like Transmission lines for Microwave Integrated Circuits*, Wiley Eastern Limited, Delhi, Dec. 1990.

Devabhaktuni, V. K., M. C. E. Yagoub, Y. Fang, J. Xu, and Q. Zhang, "Neural networks for microwave modeling: Model development issues and nonlinear modeling techniques," *International Journal of RF and Microwave Computer Aided Engineering*, vol. 11, pp. 4–21, 2001.

Haykins, S., *Neural Networks: A Comprehensive Foundation*, Prentice Hall International, NJ, ISBN: 9780139083853, 842p., 1999. http:// www.mwoffice.com.

Patnaik, A., R. K. Mishra, G. K. Patra, and S. K. Dash, "An artificial neural network model for effective dielectric constant of microstripline," *IEEE Transactions on Antennas and Propagation*, vol. 45, no. 11, pp. 1697, 1997.

Pattanayak, S., B. Choudhury, and A. Patnaik, "Characterization of planar transmission lines using ANN," *Silver Jublee conference on Communication and VLSI Design, CommV-2009*, Oct. 08-10, Vellore, India.

Raida, Z., "Modeling EM structures in neural network toolbox of Matlab," *IEEE Antennas and Propagation Magazine*, vol. 45, no. 1, pp. 46–67, 2003.

Schellenberg, J. M., "CAD models for suspended and inverted microwave microstrip," *IEEE Transactions on Microwave Theory and Technique*, vol. 43, pp. 1247–1252, June 1995.

Wheeler, H. A., "Transmission-line properties of parallel wide strips by a conformal-mapping approximation," *IEEE Transactions of Microwave Theory and Techniques*, vol. 12, pp. 280–289, May 1964.

Yildiz, C. and O. Saracogulu, "Simple models based on neural networks for suspended and inverted microstrip lines," *Microwave and Optical Technology Letters*, vol. 39, no.5, pp. 383–389, Dec. 2003.

Yildiz, C., S. Sagiroglu, and M. Turkmen, "Neural model for coplanar waveguide sandwiched between two dielectric substrates," *IEE Microwave Antennas and Propagation*, vol. 151, no.1, pp. 7–12, Feb. 2004.

Zhang, Q. J., and K. C. Gupta, *Neural Networks for RF and Microwave Design*, Artech House, ISBN: 9781580531009, 369p., 2000

Fault Detection in Antenna Arrays

Modern day communication systems and other microwave systems like sonar, radar, etc., use antenna arrays for signal acquisition. Arrays allow the designer to attain highly directional patterns that can be steered in the required directions. These arrays are typically composed of a large number of elements. Due to the presence of this large number of elements, the probability of experiencing faults in some of the elements is very high. Consequently, diagnosis of faults in a large array is a problem that antenna engineers need to tackle often. These faults—elements that do not contribute to the radiation pattern—damage the pattern by increasing the sidelobe levels. Traditionally, engineers conduct measurements in the near field of the antenna in order to pinpoint the location of these faults [Lee *et al.*, 1988; Migliore and Panariello, 2001; Bregains *et al.*, 2005]. This technique is not feasible if the antenna is mounted on a remote system like a spacecraft and human access to the system is impossible [Lord *et al.*, 1992]. Nonetheless, spacecraft antenna arrays either use test couplers or other calibration probes in the beam forming network, to detect failed elements and send the information to the ground using telemetry. These calibration probe based networks are very much complex and expensive [Bucci *et al.*, 2000]. This brings about a need to devise methods to detect faults in antenna arrays by studying the far-field radiation pattern. In this chapter, soft computing is used to meet this objective, i.e., to detect faults in antenna arrays by studying the information obtained from its far-field radiation pattern.

7.1 Preliminaries and Overview

Failure of individual elements in an antenna array results in destruction of symmetry and often causes unacceptable distortion of the radiation pattern. Locating the defective elements in large arrays is a problem that is often described using the theory of inverse scattering. A practical solution to this problem may be obtained by installing sensors in the beam forming network. However, such a sensor network is often expensive and might be susceptible to the same faults that the elements succumb to [Bucci *et al.*, 2000]. A simpler solution, therefore, is to study the far-field pattern of the antenna array and predict the location of these failures. This technique ensures independence from on-site measurements as well as disruption from normal operation of the array.

In parallel, many research groups have also proposed different techniques to compensate for faults in antenna arrays [Mailloux, 1996; Steyskal and Mailloux, 1998; Levitas *et al.*, 1999; Yeo *et al.*, 1999]. The studies in this area reveals that the degradation of side lobe level (SLL) and half power beam width (HPBW) in an antenna array depends on the number of faults and their position [Choudhury *et al.* 2013]. Various test cases of SLL degradation with respect to number of faults and fault position are described with examples of Dolph–Chebyshev array. Further, the recoveries of radiation pattern analysis with failed elements are also discussed by Acharya *et al.* 2014. In a test case of 32 element Dolph–Chebyshev antenna array, with SLL –30 dB, single element fault takes the damaged SLL to –25 dB and can be recovered to –29.99 dB. Similarly, with two element fault, it has been reported that the damaged SLL goes to –22.93dB and can be recovered to –29.83 dB.

These techniques have found application in many wireless systems like MIMO (Multiple Input, Multiple Output), which enhances the capacity of the wireless channel by taking advantage of the spatial location of different antennas. The principle behind these techniques is roughly the same and is based on the fact that the excitations for beam forming can be controlled remotely. However, the first step involves the location and total number of the faults in the array.

As mentioned earlier, ideally, the problem would be solved if far-field measurements can provide engineers with the location of these faults. This would eliminate the requirement for near-field diagnostic tools like those that have been proposed in literature [Lee *et al.*, 1988; Migliore and Panariello, 2001; Bregains *et al.*, 2005; Gattoufi *et al.*, 1997]. The challenge to far-field diagnostic procedures is the inherent asymmetrical nature of the problem. As a result, finding out the solution becomes a tedious task. In this chapter, a method to overcome this challenge is presented using soft computing techniques. The proposed technique will eliminate the need for near-field measurements, a task that is tough in remote antenna array installations like spacecraft, etc.

The objectives of this chapter are two-fold. First, an effective soft computing method is developed that can be used to find the positions and number of faults in an antenna array using the information obtained from its far-field radiation pattern measurements. Secondly, a time analysis of the procedure is performed in order to check the feasibility of this technique for implementation in on-line systems.

The problem at hand is solved using three separate soft computing techniques in order to demonstrate the effectiveness of the proposed technique. Initially, the task of array fault finding is converted to a regression problem and neural network is used to arrive at a solution. Following this, *particle swarm optimization* is applied to the same problem for the identification of the location of the faults. Later, the implementation of *bacterial foraging optimization* is demonstrated for the same problem.

7.2 Artificial Neural Network for Array Fault Detection

A neural network is an artificial system that mimics the behaviour of actual neurons and memory present in living beings. Neural networks are simplified mathematical models and have been regarded as extensions to conventional data processing technique. An insight into the utility of this tool has been provided in Chapter 2. The neural network described in this chapter relies extensively on the theory presented in Chapter 2.

Application of neural networks in antenna engineering has been evolving rapidly with several research groups making significant contributions [Mishra *et al.*, 2003; Patnaik *et al.*, 2005, Christodoulou *et al.*, 2000, Zhang *et al.*, 2000]. From the introduction to neural networks given in Chapter 2, the main reasons for the popularity of application of the algorithm to antenna engineering are the following:

- **Simplification of Non-Linear Problems:** Typical problems in antenna engineering, viz. design and analysis, adaptive beamforming, estimation of direction of arrival, etc., have a non-linear relationship between the outputs and inputs. Multilayer feed-forward neural networks possess the capability of solving these non-linear problems and such a network may be considered as a *universal approximator*; one that provides approximate solutions to non-linear problems [Hornik *et al.*, 1989].

- **Ability to develop CAD packages:** Conventional design software employs intensive numerical methods like full-wave MoM, FDTD, FEM, etc. The simulations using these tools are time consuming. On the other hand, neural networks have the capability to reduce the total computation time due to the inherent parallelism and connected structure of the algorithm. It is conceded at this stage that neural networks consume a lot of time during the training phase. However, this might be overlooked considering that the results during the implementation phase are nearly instantaneous. Therefore, neural networks have great potential in the design of CAD packages.

- **Reduction in Mathematical Complexity:** Neural networks use complex mathematical calculations to train the network. However, once trained, the neural network does not need the mathematical computations in order to arrive at the solution to the problem at hand. Therefore, the computational complexity at the implementation stage is significantly reduced. Many researchers have also used neural networks in conjunction with conventional Physical/EM solvers in order to reduce the overall complexity of the simulations [Mishra 2002].

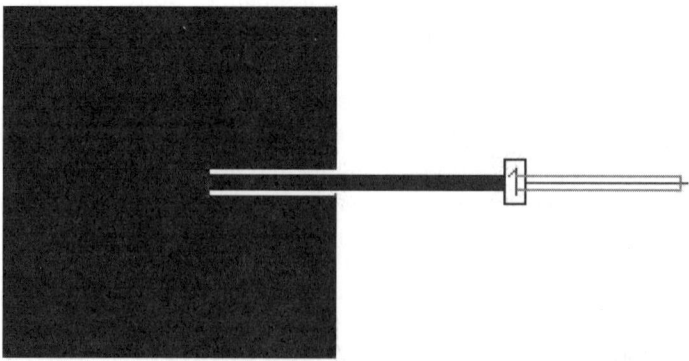

Fig 7.1 Schematic diagram of the single element of the array

Taking all these properties into consideration, this section focuses on the technique of identifying faults in antenna arrays using neural networks. This task was modelled as a regression problem and the function mapping capabilities of feed-forward neural networks is used to solve it. The neural network is capable of forming a mapping between the degraded pattern and the position of the faults. In this chapter, this mapping is achieved using *multi-layer perceptron* (MLP) trained in backpropagation mode as discussed in Chapter 2.

7.2.1 Antenna array design

A 16-element, rectangular patch microstrip array with elements arranged in a linear fashion was chosen for this illustration. All elements were designed to resonate at 1.9 GHz as shown in Fig. 7.1. The dimensions of these elements are as follows: patch length = 1.512", patch width = 1.5", inset width = 0.115", inset depth = 0.452", strip width = 0.06", feed line length = 0.75". A substrate of dielectric constant 4.4 and thickness 0.031" was chosen for the design. These rectangular patch elements were arranged with a spacing of 0.3 λ in order to generate the array (Fig. 7.2). The various dimensions of the array are as follows: total length of transmission line = 1.2", horizontal length of the bend = 0.4", vertical length of the bend = 0.6", total length of power combiner = 0.3". The highly directional nature of this array is shown in its 3-D radiation pattern shown in Fig. 7.3.

This highly directional pattern array pattern is disrupted when a fault occurs in one or more array element. Consequently, an increase in the sidelobe levels and ripples in the power pattern are observed. The magnitude of the effect of these faults depends on the position and number of faults in the array. For example, Fig. 7.4 (a) shows the power pattern obtained for a fault at the fifth element.

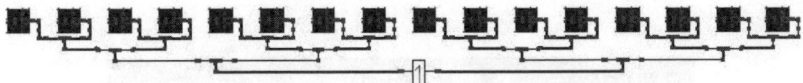

Fig 7.2 Schematic diagram of the 16-element linear microstrip patch array

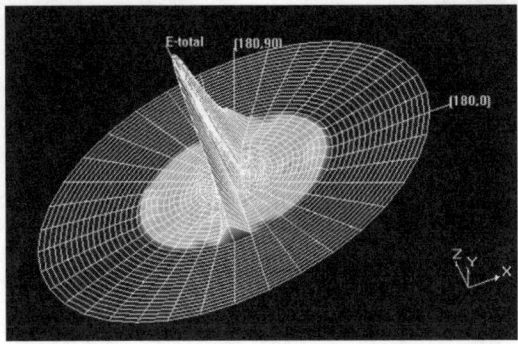

Fig 7.3 3D radiation pattern of the 16-element linear microstrip array without any faults

Figure 7.4 (b) shows the power pattern obtained when the array experiences two faults at the 7th and 9th elements. In this section, a fault has to be assumed to be completely non-radiating in nature. It must be conceded that in practice, the faults might radiate partially as well. Presence of these partially faulty elements also degrades the overall pattern of the array. Increase in the number of these arbitrarily occurring faults led to further degradation of the pattern.

The task of mapping the distorted radiation pattern obtained from far-field measurements to the location and number of faults in the array is carried out by a multi-layer perceptron (MLP) neural network. The neural network is designed to take the far-field radiation pattern as the input and provide the location of faults as the output as shown in Fig. 7.5.

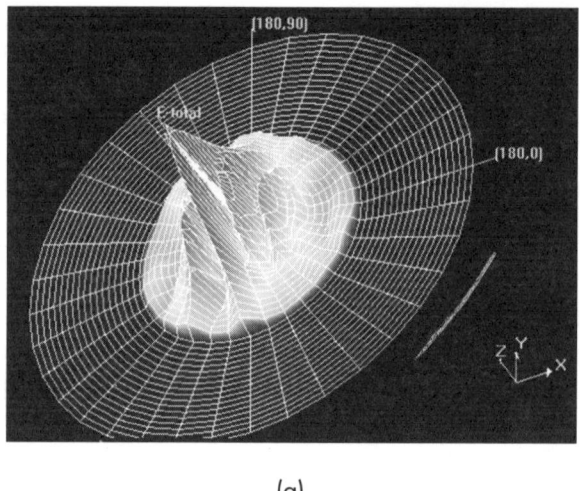

(a)

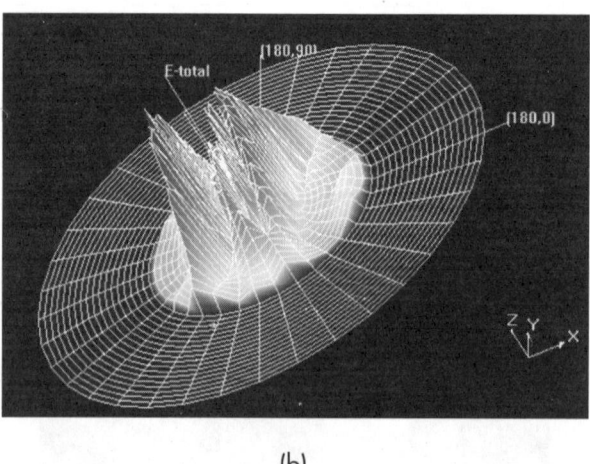

(b)

Fig 7.4 3D radiation patterns of the 16-element linear array with (a) single (5th) element faults and (b) double (7th and 9th) element fault

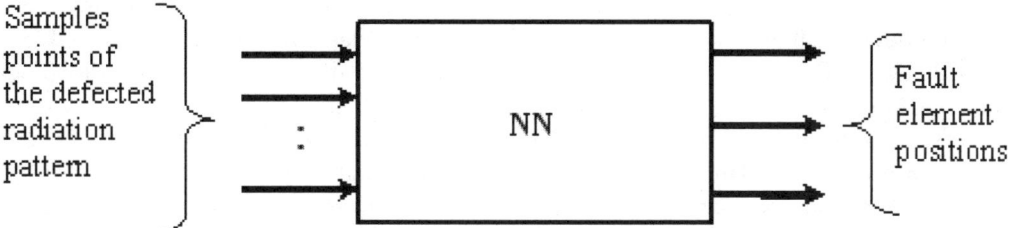

Fig 7.5 Input and output parameters of the neural network

In this chapter, actual field data has not been used as the input. Instead, the input to the neural network is considered to be the data from a faulty array simulated using IE3D. Since the aim of this chapter is to demonstrate the effectiveness of neural networks in fault detection in antenna arrays, simulated input data was considered to be sufficient. The 3-D radiation pattern obtained from IE3D is converted to a 2-D pattern by considering only the $\phi = 0°$ principal plane. This 2-D plot contains all the relevant information about the array.

For a fault-less array, the 2-D radiation pattern in the $\phi = 0°$ plane is plotted in Fig. 7.6. In this chapter, the maximum number of faults in the array has been restricted to three. The methodology can be extended to include more faults. However, the probability of encountering a large number of faults is low.

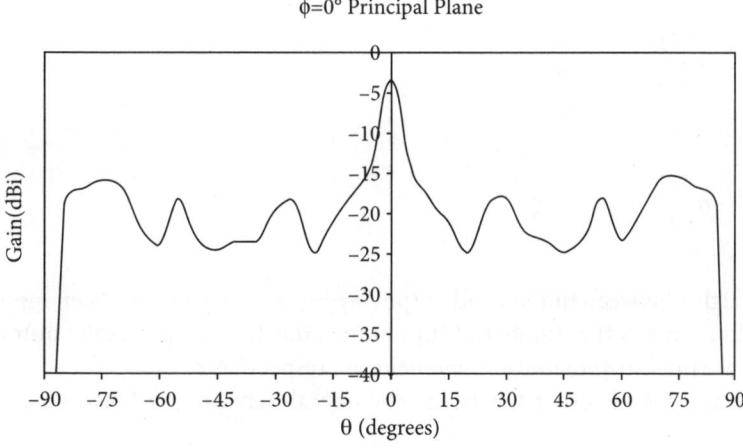

Fig 7.6 Radiation pattern of a 16-element linear microstrip array in the absence of faulty elements

7.2.2 ANN implementation

There are three basic steps for implementation of any neural network algorithm, viz. *data generation*, *network training* and *testing*. A brief about these steps for the current problem is given below.

Data Generation: In order to begin the training of the neural network, radiation patterns for randomly located single element, two-element and three-element faults are generated. The faults are assumed to be non-radiative. The 3-D radiation pattern obtained from simulation using IE3D is converted to a 2-D pattern by observing only the $\varphi = 0°$ plane.

The study of array theory reveals that the discrete Fourier transform of excitation coefficients gives the array factor [Balanis, 2005]. Therefore, the fault analysis problem is akin to deriving a cost function from the corresponding modules of the Fourier transform and is achieved by sampling the pattern at equal intervals using the sampling theorem. Therefore, this 2-D pattern is then sampled at an interval of 5° and 37 samples from each pattern are taken as the input set to the neural network. Since the output corresponds to the location of the faults (with a maximum of 3 faults), the output is considered to be a three element vector with each element ranging from 0 to 16.

Network Training: As mentioned in the earlier section, the MLP neural networks have the capability of arriving at the approximate solution to generic equations [Scarselli and Tsoi, 1998]. MLP trained in error backpropagation mode have been used by researchers in antenna engineering [Scarselli and Tsoi, 1998; Mishra, and Patnaik, 1998; Patnaik et al., 2005]. For every iteration, the network adjusts the thresholds and weights using the following equations:

$$v_{ho}^{k+1} = v_{hj}^{k} - \eta \frac{\partial E^k}{\partial v_{ho}} + \alpha(v_{ho}^{k} - v_{ho}^{k-1}) \tag{7.1}$$

$$w_{ih}^{k+1} = w_{ih}^{k} - \eta \frac{\partial E^k}{\partial w_{ih}} + \alpha(w_{ih}^{k} - w_{ih}^{k-1}) \tag{7.2}$$

$$\theta_{h}^{k+1} = \theta_{h}^{k} - \eta \frac{\partial E^k}{\partial \theta_{h}} + \alpha(\theta_{h}^{k} - \theta_{h}^{k-1}) \tag{7.3}$$

where, v_{ho} is weights between hidden and output layers; w_{ih} is weights between input and hidden layers; $i = 1, 2, ..., n$; n is the number of input units; $o = 1, 2, ..., p$; p is the number of output units; η and α are the *learning rate* and *momentum*, respectively.

Training minimizes the error E between the neural network predicted y_o, and the desired outputs d_{ko}

$$E = \sum_{k=1}^{N} E^k = \sum_{k=1}^{N} \left[\frac{1}{2} \sum_{o=1}^{P} (y_o - d_{ko})^2 \right] \tag{7.4}$$

where, $k = 1, 2, ..., N$; N is the total number of samples considered for training. The sigmoidal transfer function is usually employed for the hidden layer neurons. For the outputs, the transfer functions are considered to be linear.

The neural network acquires knowledge from samples of input and output pairs, i.e., a_k and b_k, $k = 1, 2, ..., N$. In this example, radiation patterns for 697 fault combinations are simulated.

Since the network size is large compared to the total number of training patterns, regularization method is adopted to improve network generalization and avoid over-fitting. In this method, the feed forward network performance function is modified by introducing a mean square error term consisting of average weights (mean of the sum of squares of the weights) and biases. The modified performance function is given by

$$E = \gamma \left[\frac{1}{N} \sum_{i=1}^{N} (t_i - a_i)^2 \right] + (1 - \gamma) \left[\frac{1}{n} \sum_{j=1}^{n} w_j^2 \right] \tag{7.5}$$

where, γ is the performance ratio, a_i is the output of the network (y), t_i is the target output (d), and n is the total number of weights and biases. For the implementation of the regularization procedure, the inputs and outputs are prepossessed to lie within the range [0,1].

The optimal values of the training parameters of the neural network (described in Chapter 2) are problem-dependent. As a result, these values are often found out using a trial and error method. Table 7.1 shows the training parameters and their values considered for the present network.

7.2.3 Results

The trained neural network is then subjected to performance testing. Sampled radiation pattern of arrays with different faults are considered to be the input to the neural network. Figures 7.7, 7.8, and 7.9 show the radiation pattern obtained for a single fault, two faults and three faults, respectively. Each element in the three element vector output of the network is rounded off in order to arrive at the location of the faults. For example, a raw output of 0.602 is rounded off to 0.6 representing a fault at the 6th position. The outputs of the network with the corresponding post-processing values for an array with three faults are given in the inset box in Fig. 7.9 and matches well with the simulation results.

Table 7.1 Network/Training parameters of the NN developed for fault finding in linear array

Parameters	Value
Neurons in the input layer	37
Neurons in the output layer	3
Number of hidden layers	1
Neurons in the hidden layer	15
Learning rate (η)	0.05
Performance ratio (γ)	0.7
Training tolerance	1×10^{-2}
Training time (2GB RAM, Intel PC)	15 minutes

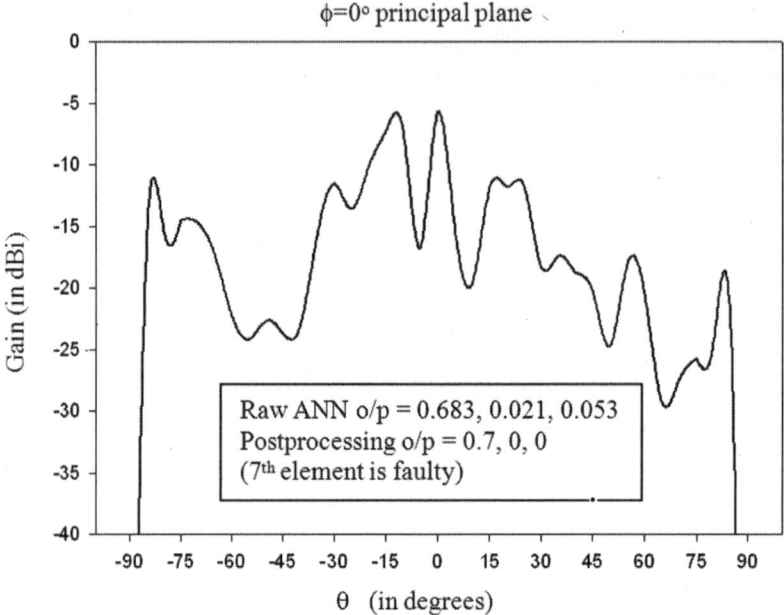

Fig 7.7   Pattern for single (7th) element fault in the linear array with the corresponding NN output in the inset

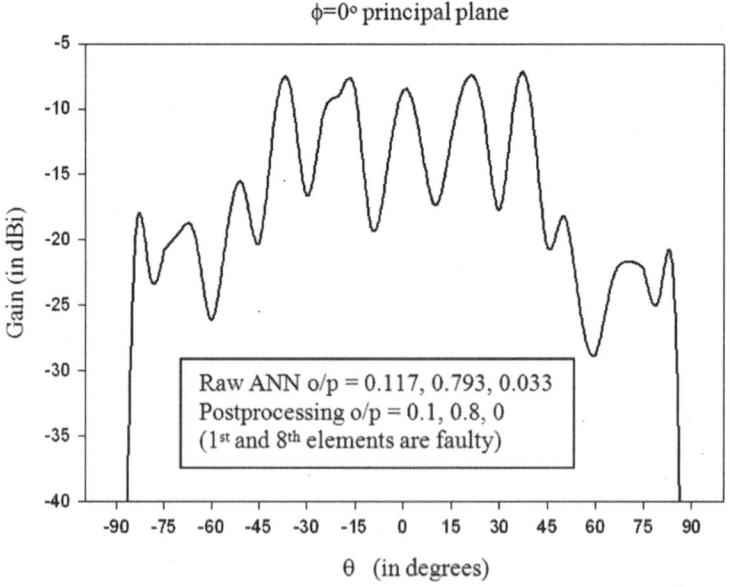

Fig 7.8   Pattern for double (1st and 8th) element fault in the linear array with the corresponding NN output in the inset

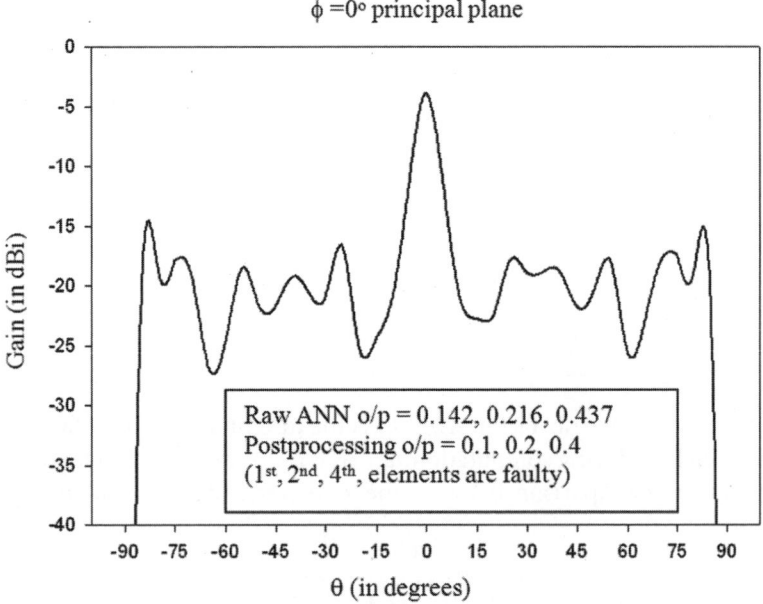

Fig 7.9 Pattern for triple (1st, 2nd, and 4th) element fault in the linear array with the corresponding NN output in the inset

7.3   PSO for Array Fault Detection

In the previous section, location of faults in an antenna array using the information available in its far-field radiation pattern was carried out using neural networks. The number of elements in the array used for the demonstration of this technique was 16. As the number of elements is increased, the data required to train the neural network increases significantly. This reduces the speed of the algorithm and at the same time increases the memory requirement. As a result, in this section, the problem is solved using *particle swarm optimization* instead of other conventional optimization techniques [Robinson and Rahmat–Samii, 2004]. While using PSO, the input radiation pattern is compared to the radiation pattern obtained for an array with randomly generated faults.

In the Section 7.2.2, the reason for taking equidistant samples of 2-D radiation pattern was presented. However, obtaining equidistant points with sufficient radiated power is not always feasible. In this chapter, the possibility of detecting faults by random sampling of the far-field radiation pattern is discussed. Further, the algorithms have been extended to include complete and partial faults, i.e., completely non-radiating faults as well as partially radiating faults.

7.3.1 PSO implementation

For an N-element linear, non-uniform array with progressive phase excitation, the array factor is given by [Balanis, 2005]

$$AF(\theta) = \sum_{n=1}^{N} a_n e^{j(n-1)(kd\cos\theta + \beta)} \tag{7.6}$$

where, a_n's are the non-uniform amplitude excitation of elements, d is the spacing between the elements and β is the progressive phase shift.

The pattern of the faulty array is generated using Eq. (7.6) by equating amplitude excitation to zero for a failed element, and half to denote a partially defective element. The following fitness function is then computed

$$I = \sum_{k=1}^{M} \left[\left| AF_d(\theta_k) - AF_o(\theta_k) \right| \right]^2 \tag{7.7}$$

where, AF_d is the measured array factor, $AF_o(\theta_k)$ is the instantaneous array factor at k^{th} sample point obtained from PSO during computation, and M is the number of sample points considered on the array pattern. Comparison between the radiation pattern from the optimizer and

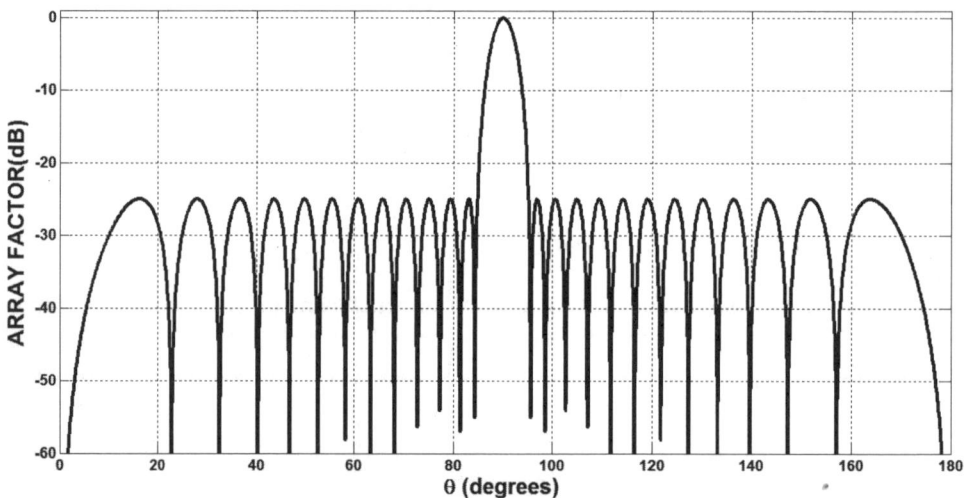

Fig 7.10 Array factor for 26-element linear array with −25dB sidelobe level

the radiation pattern of the array under consideration provides insight into the nature, number and location of the faults.

In this section, a 26 element broadside Chebyshev array is considered to be the antenna under test. The inter-element spacing in the array is taken to be $\lambda/2$. The excitation amplitudes to each of the elements are determined in order to obtain a −25 dB sidelobe level [Balanis, 2005] as shown in Fig. 7.10. The normalized amplitude excitation of each array is given in Fig. 7.11.

Complete and partial faults are introduced randomly in the Chebyshev array by making the excitation of some elements equal to zero, or half of the maximum respectively. Equation (7.7) was then minimized using particle swarm optimization, the algorithm for which is given in Chapter 2.

Consider an example in which the Chebyshev array has succumbed to four faults: a complete fault at the 6th element and partial faults at the 9th, 12th, and 20th elements. The radiation pattern of this faulty array is shown in Fig. 7.12. The distortion in the pattern is clearly seen from the fact the sidelobe level has increased by approximately 12 dB. Initially, this radiation pattern is sampled 35 times (range: 0° $\leq \theta \leq$ 180° for every 5°). These points are used to construct the fitness function for the PSO algorithm.

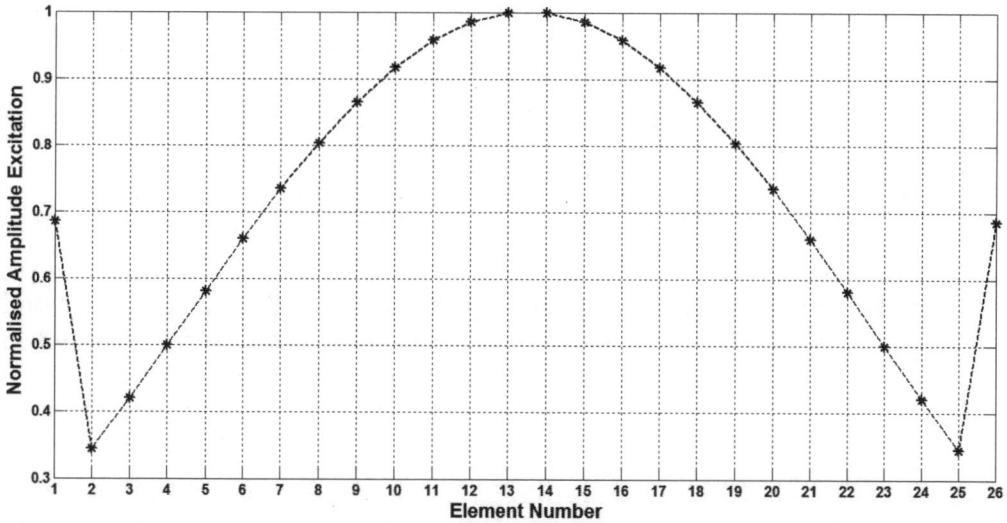

Fig 7.11 Normalized amplitude excitations of array

The PSO is then run with the set of amplitude excitations to the array acting as the dimensional parameters. The fitness function is minimized as the number of iterations increased as shown in Fig. 7.13. The PSO parameters used in the optimization are given in Table 7.2.

Table 7.2 PSO parameters

Parameter	Value
Number of particles	30
Inertial weight (w)	Linearly damped with iterations from 0.9 to 0.4
Cognitive parameter(c_1)	2
Social parameter(c_2)	2
Random function (rand(*range*))	Range [0,1]

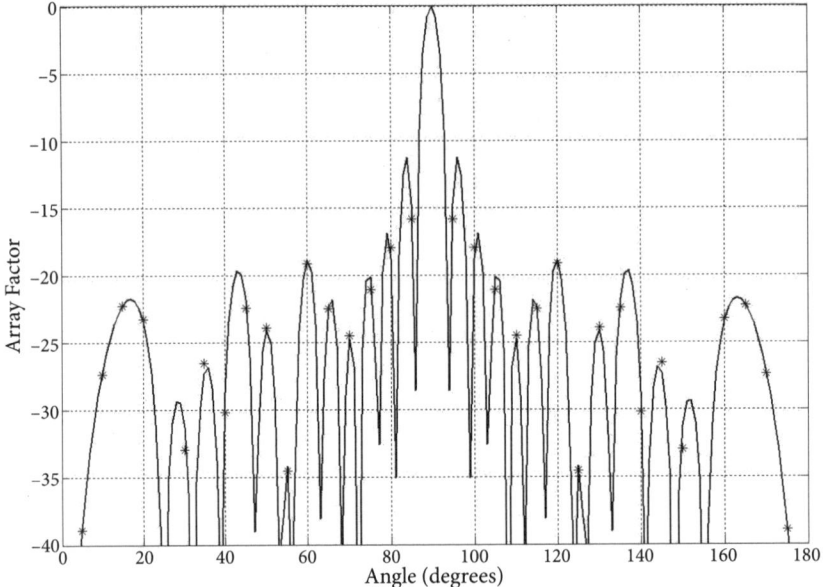

Fig 7.12   Defected array pattern with faults in 9th (50%), 12th (50%) and 20th (50%), and 6th (100%) elements. The sample points taken for framing the cost function is marked with (*) symbols

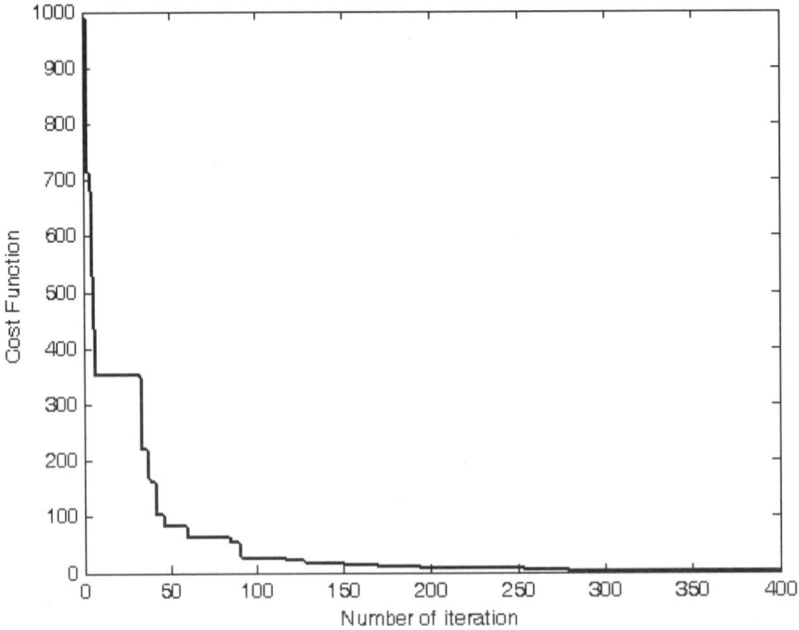

Fig 7.13   Error performance of PSO

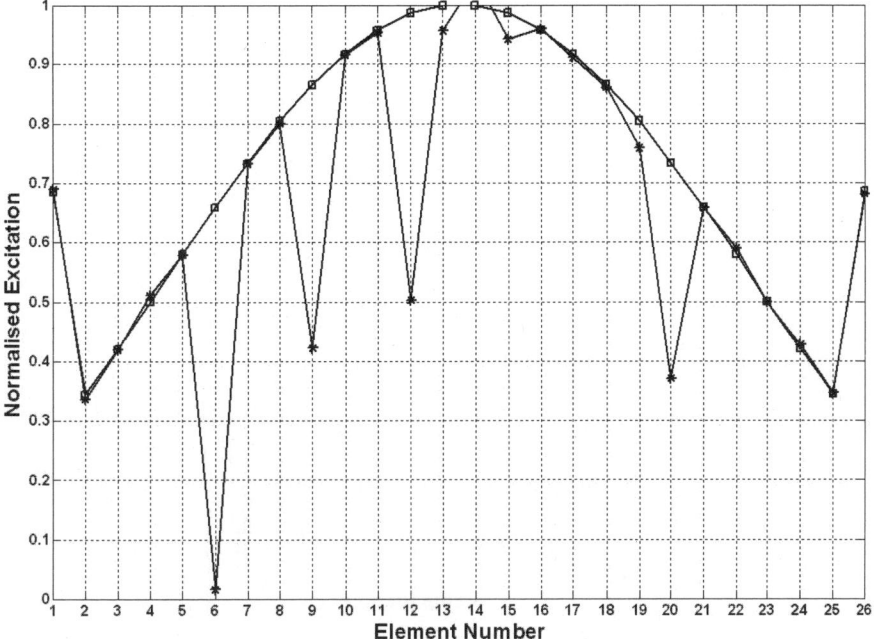

Fig 7.14 Performance of PSO with 35 sample points

Completion of PSO based on a predefined error tolerance gives a set of array element amplitude excitation values. A comparison between the amplitude excitations given by the PSO solution with that of the original array (without any defect) gives an idea about the nature and location of faults in the array. This comparison is shown in Fig. 7.14, i.e., the PSO algorithm has determined that partial faults are present at the 9th, 12th, and 10th positions and a complete fault is present at the 17th position.

7.3.2 Results and discussion

As mentioned earlier, the amplitude excitations of all the elements are taken together to form the set of dimensions for the PSO algorithm. The values in this set are iteratively varied so that the fitness function becomes minimum (the fitness function being defined in terms of the sampled radiation pattern of the faulty array). Comparison of the amplitude excitations obtained from PSO with that of the array without any faults gives an idea about the number and location of faults. Hence, the crux of the problem is the determination of the ideal number of samples. In this regard, the same fault situation is tested with 50% the number of sample points in the previous section, i.e., 18 sample points as shown in Fig. 7.15.

With 18 sample points, the PSO was able to correctly determine the location of the faults. The comparison between the amplitude excitations of the original array without faults and the amplitude excitations of the faulty array obtained from PSO is shown in Fig. 7.16. The position and nature of the faults are seen clearly from the figure. The total number of sample

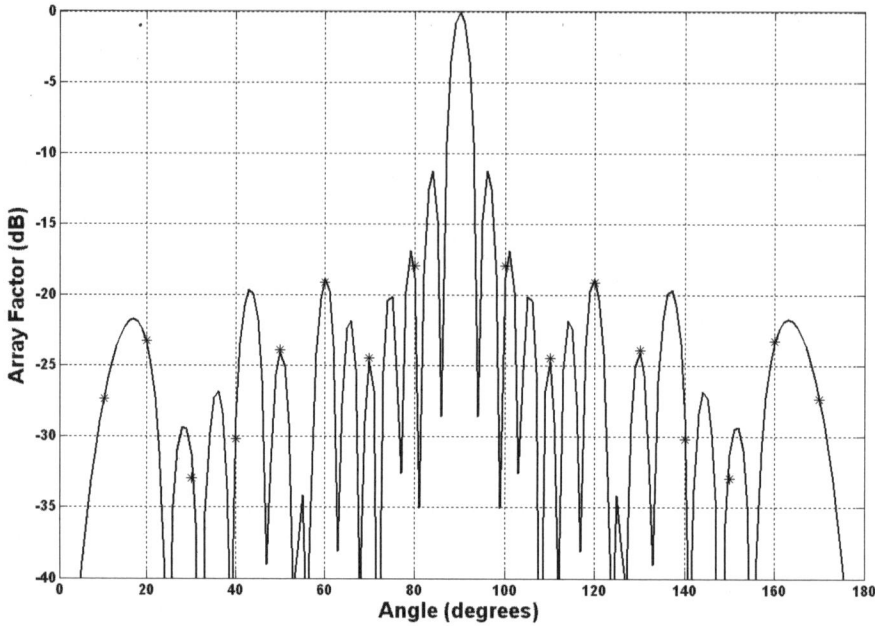

Fig 7.15 18 sample points on the array factor plot of the same defected array

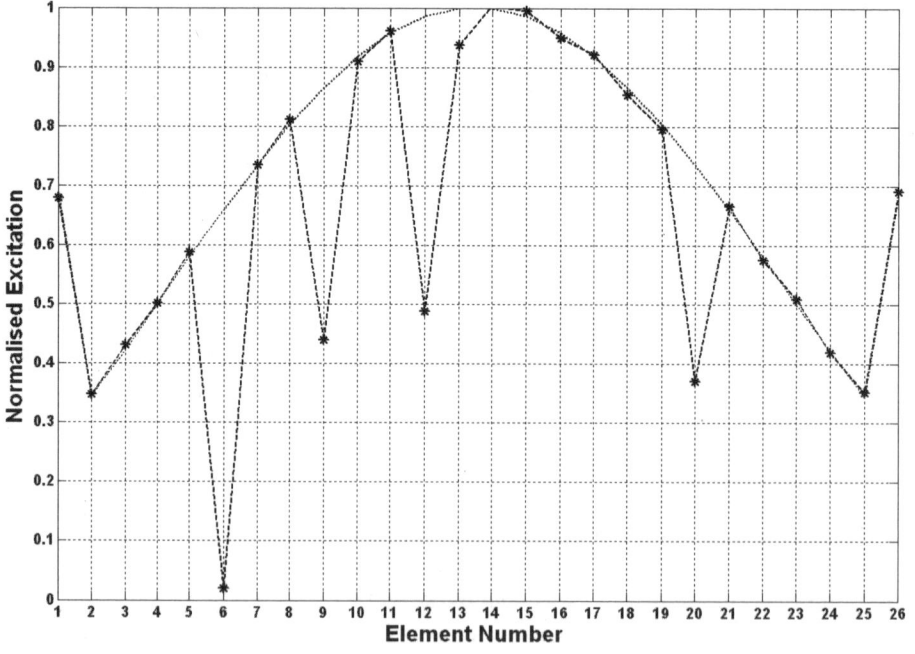

Fig 7.16 PSO performance (18 sample points)

points is then decreased and the effect on the efficiency of the PSO is studied. It was noted from the observations that locating the faults for the range $0° \leq \theta \leq 180°$ become impossible, if the number of samples is less than 12. Further, an attempt is made to see if random samples would suffice in the PSO implementation. The biggest reason for checking this test case is that from a practical point of view, sampling the far-field radiation pattern at equidistant points is not always possible. Therefore, the radiation pattern is sampled at random points and used to construct the fitness function. Fifteen samples points are taken in this example, at $\theta = 13°, 34°, 42°, 56°, 69°, 88°, 98°, 102°, 126°, 131°, 147°, 167°, 171°$ as shown in Fig. 7.17. On simulation, it is observed that PSO could detect the location of faults even with a randomly sampled radiation pattern. The result of this simulation is shown in Fig. 7.18.

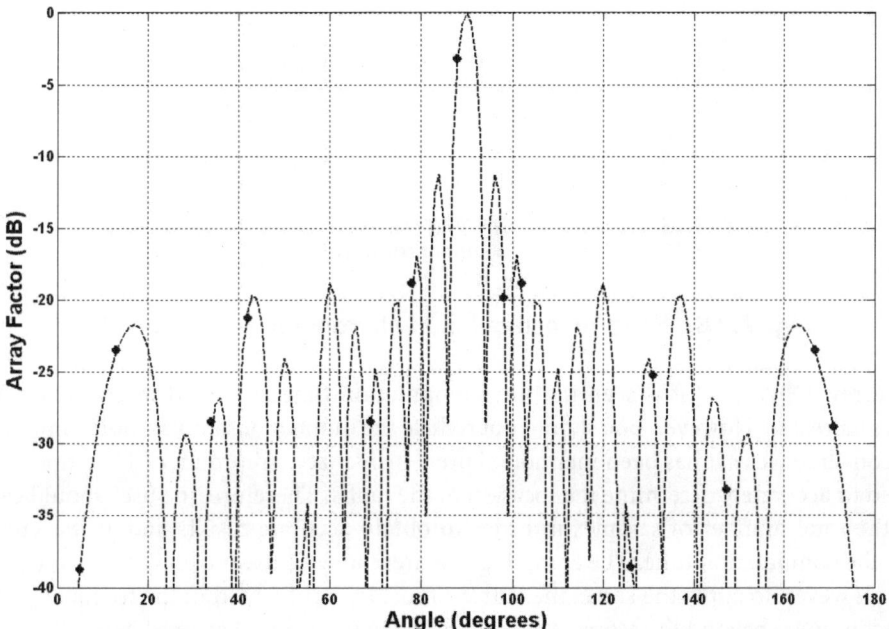

Fig 7.17 Sample points taken at random locations on the array factor plot of the same defected array

The numerical values for amplitude excitations for each element of the array obtained from the PSO algorithm for the three cases, viz. with 18 equidistant samples, 35 equidistant samples, and 15 random samples, are given in Table 7.3.

The faults are clearly detected when the variation between the amplitude excitation of the element under consideration and that of the original Chebyshev array is large. In Table 7.3, the location of the faults has been marked in bold.

The average computation time for this algorithm is determined by averaging the computation time required to detect faults at five different locations separately. Tables 7.4–7.6 show the average computation time for one, two and three faults, respectively. Table 7.7 shows the average computation time taken when a combination of partial and complete faults are present

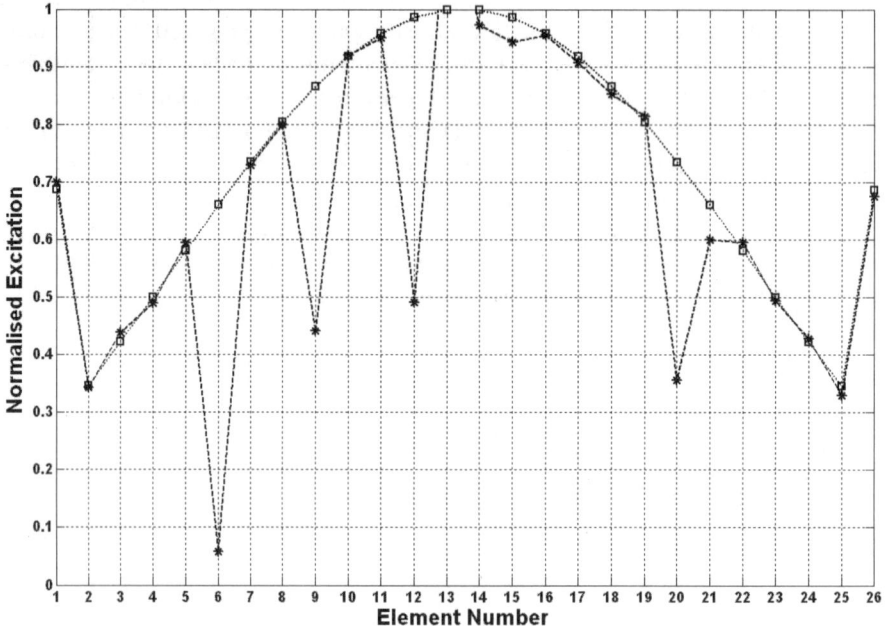

Fig. 7.18 Performance of PSO with random sample points

in the array. Obviously, it is seen that the computation time increased when the number of samples increased. However, one should not reduce the number of samples indiscriminately for faster convergence as it has been mentioned previously that a minimum of 12 sample points is required to accurately determine the location of the faults. Therefore, the user should carefully select the total number of samples in order to obtain accurate results and at the same time reduce the computation time. The Sampling Theorem may be used to solve this issue [Proakis, 2001]. However, to apply the same, the pattern must be represented in the frequency domain. This technique is beyond the scope of this book and hence is not discussed here.

Table 7.3 Element excitations with different number of sample points for the array with faults at 9th (partial), 12th (partial), 20th (partial) and 6th (complete) positions

Element Number	Chebyshev Excitation	18 Samples	35 Samples	Random Samples
1	0.6859	0.6798	0.6898	0.7004
2	0.3451	0.3478	0.3367	0.3434
3	0.4215	0.4304	0.4198	0.4379
4	0.5007	0.5013	0.5106	0.4894
5	0.5808	0.5876	0.5796	0.5945
6	**0.6596**	**0.0204**	**0.0167**	**0.0580**
7	0.7348	0.7356	0.7321	0.7287
8	0.8043	0.8112	0.8011	0.7989

9	**0.8659**	**0.4398**	**0.4224**	**0.4414**
10	0.9178	0.9103	0.9167	0.9205
11	0.9582	0.9605	0.9545	0.9501
12	**0.9859**	**0.4887**	**0.5023**	**0.4905**
13	1.000	0.9381	0.9566	0.9789
14	1.000	0.9905	0.9989	0.9724
15	0.9859	0.9946	0.943	0.9432
16	0.9582	0.9488	0.9601	0.9553
17	0.9178	0.9209	0.9123	0.9067
18	0.8659	0.8534	0.8612	0.8523
19	0.8043	0.7956	0.7599	0.8145
20	**0.7348**	**0.3715**	**0.3698**	**0.3567**
21	0.6596	0.6657	0.6578	0.5992
22	0.5808	0.5743	0.5899	0.5956
23	0.5007	0.5087	0.5004	0.4934
24	0.4215	0.4187	0.4279	0.4278
25	0.3451	0.3512	0.3467	0.3301
26	0.6859	0.6904	0.6823	0.6745

Table 7.4 Time analysis for computation of one defective element (5 random configurations)

Failed element position	Time (s) (18 samples)		Time (s) (35 samples)	
1	1.52		2.62	
3	1.49		2.54	
13	1.59	Average Time: **1.514**	2.68	Average Time: **2.632**
18	1.46		2.73	
25	1.52		2.59	

Table 7.5 Time analysis for computation of two defective elements (5 random configurations)

Failed element position	Time (s) (18 samples)		Time (s) (35 samples)	
6, 12	2.1		4.54	
9, 17	2.21		4.68	
23, 25	2.12	Average Time: **2.164**	4.32	Average Time: **4.486**
1, 13	2.16		4.39	
5, 21	2.23		4.50	

Table 7.6 Time analysis for computation of three defective elements (5 random configurations)

Failed element position	Time (s) (18 samples)		Time (s) (35 samples)	
3, 10, 16	2.56		4.45	
1, 17, 21	2.67		4.91	
6, 12, 19	3.17	Average Time (s) **2.722**	5.87	Average Time (s) **5.068**
11, 14, 23	2.78		4.14	
4, 15, 25	2.43		5.97	

Table 7.7 Time analysis for computation of combination of complete and partial defective elements

Failed element position	Time (s) (18 samples)		Time (s) (35 samples)	
4 and 16(100%), 20(50%)	3.78		4.67	
6 and 23(50%), 15(100%)	2.34		5.12	
10 and 25(100%), 1 (50%)	4.01	Average Time: **3.09**	4.96	Average Time: **5.018**
7 and 15 (100%), 24(50%)	3.33		5.33	
2 and 19(50%), 17(100%)	2.01		5.01	

Further, it may be observed that the computation time increases as the number of faults increases. It should also be mentioned here that as the number of faults increases further, it becomes tougher to handle the problem, and increasing the number of samples may provide better results.

7.4   BFO for Array Fault Finding

In this section, the determination of the location of faults in the array is carried out using the *bacterial foraging algorithm* (BFO). The algorithm has been discussed in detail in Chapter 2. The 26-element Chebyshev array used in Section 7.3 has also been used here.

The implementation of bacterial foraging algorithm for microwave applications is a relatively new field. First introduced by Passino in 2002, the algorithm emulates the food foraging behaviour of *E. coli* in human intestines. The group of bacteria moves in search of rich nutrients and away from noxious matter. Using the process of natural selection, bacteria with poor foraging abilities are eliminated. This strategy takes place through four steps: *chemotaxis*, *swarming*, *reproduction* and *elimination–dispersion*. The definition of each of these terms is given in Chapter 2.

7.4.1 BFO implementation

The values of the different BFO parameters are given in Table 7.8. Faults in the antenna array are generated in the same manner as given in Section 7.3. In order to compare the performance of BFO with respect to PSO, the same faults are generated during the implementation of the BFO algorithm. 35 sample points are taken for the simulation. The comparison between the BFO generated amplitude excitation and the original array is given in Fig. 7.19. The variation of the cost function with respect to the number of iterations is given in Fig. 7.20.

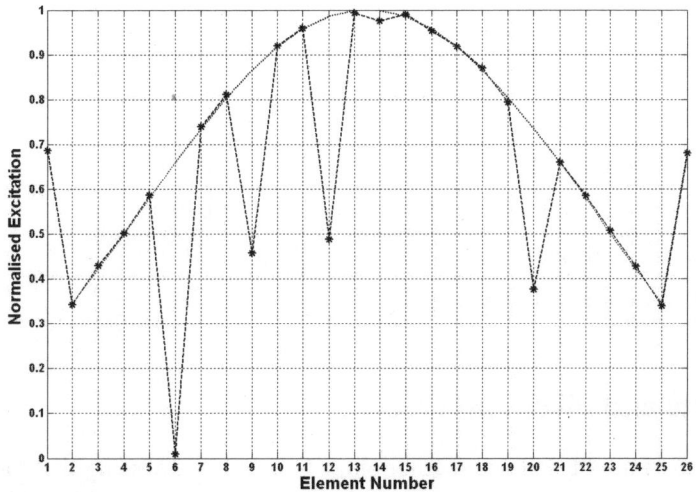

Fig 7.19 BFO performance plot (35 sample points)

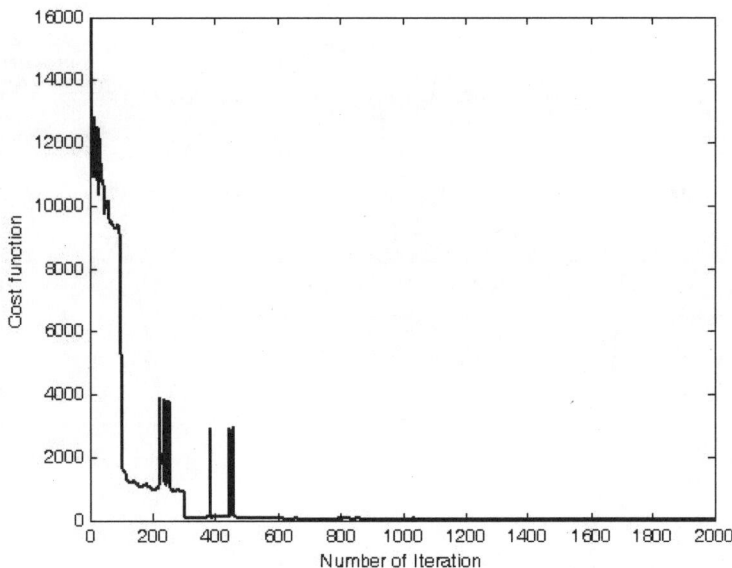

Fig 7.20 Error performances of the bacteria foraging algorithm

It can be inferred from Fig. 7.13 and Fig. 7.20, that the PSO takes lesser number of iterations to reach the same level of error tolerance. Further, the initial error in BFO is much larger than the one seen in when the problem is solved using PSO. However, this observation is dependent on the initialization of the values in the set of variables.

Table 7.8 BFO parameters

Parameters	Value
Number of Bacteria (S)	30
Swimming length (N_s)	50
Number of steps (chemotactic) N_c ($N_c > N_s$)	100
Number of reproduction (N_{re})	10
Number of elimination- dispersal events (N_{ed})	2
Probability of elimination and dispersal (P_{ed})	0.25

7.4.2 Results and discussion

The performance of the BFO based methodology is tested by reducing the number of sample points and by choosing random points instead of equidistant points. It is observed that these changes do not affect the efficiency of the algorithm. The result of the simulation for 18 sample points is given in Fig. 7.21 and the same for random sample points is given in Fig. 7.22.

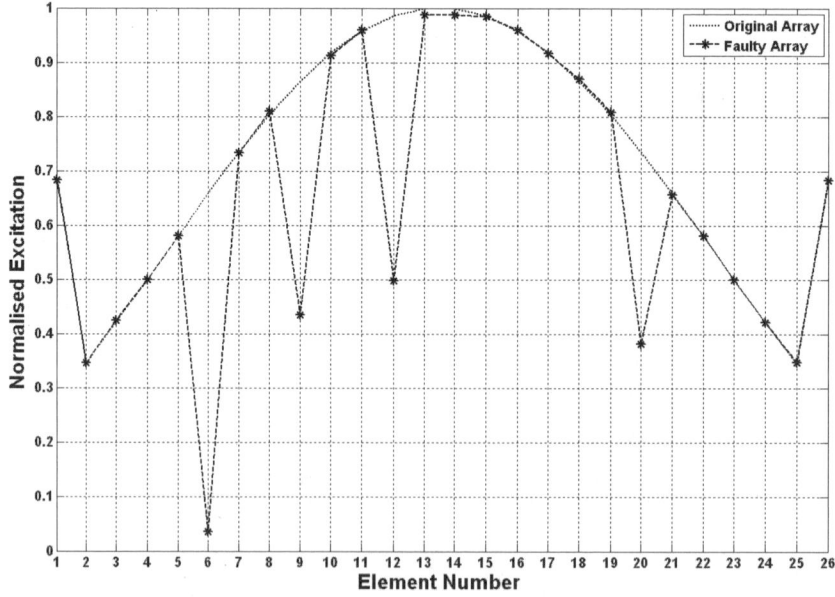

Fig 7.21 BFO performance of 26 element array (18 sample points)

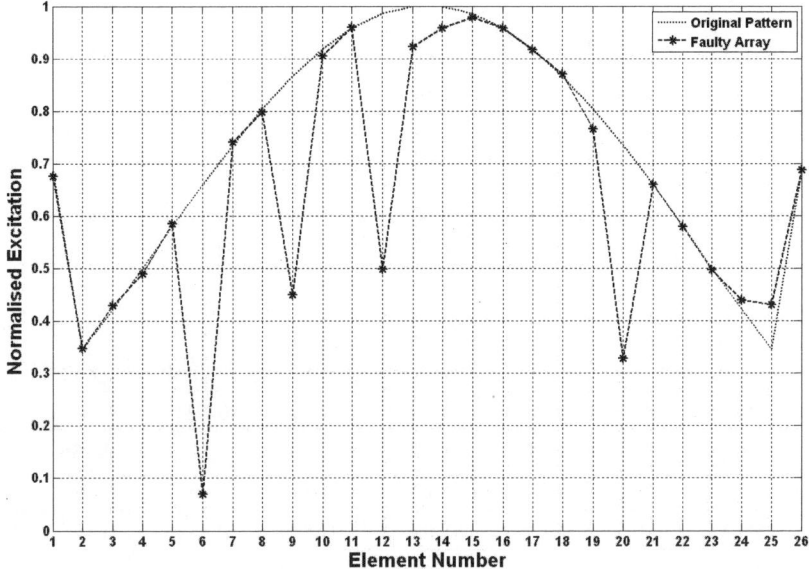

Fig 7.22 BFO performance with random sample points

Element excitations obtained using by implementing the BFO algorithm for different number of sample points is given in Table 7.9 alongside the results obtained from the PSO based methodology. While minor differences in the computed amplitude excitations are observed depending on the methodology used, these differences do not affect the final result i.e. the location of the faults.

Table 7.9 Element excitations with different number of sample points for the array with faults at 9th (partial), 12th (partial), 20th (partial) and 6th (complete) positions

Element Number	Chebyshev Excitations	18 samples		35 Samples		Random sample points	
		PSO	BFO	PSO	BFO	PSO	BFO
1	0.6859	0.6798	0.6832	0.6898	0.6867	0.7004	0.6767
2	0.3451	0.3478	0.3467	0.3367	0.3423	0.3434	0.3478
3	0.4215	0.4304	0.4245	0.4198	0.4288	0.4379	0.4290
4	0.5007	0.5013	0.4998	0.5106	0.5012	0.4894	0.4904
5	0.5808	0.5876	0.5812	0.5796	0.5870	0.5945	0.5845
6	**0.6596**	**0.0204**	**0.0356**	**0.0167**	**0.0099**	**0.0580**	**0.0698**
7	0.7348	0.7356	0.7332	0.7321	0.7399	0.7287	0.7403
8	0.8043	0.8112	0.8095	0.8011	0.8106	0.7989	0.7987

9	**0.8659**	**0.4398**	**0.4356**	**0.4224**	**0.4576**	**0.4414**	**0.4498**
10	0.9178	0.9103	0.9132	0.9167	0.9201	0.9205	0.9056
11	0.9582	0.9605	0.9598	0.9545	0.9600	0.9501	0.9597
12	**0.9859**	**0.4887**	**0.4992**	**0.5023**	**0.4879**	**0.4905**	**0.4992**
13	1.000	0.9381	0.9873	0.9566	0.9934	0.9789	0.9228
14	1.000	0.9905	0.9876	0.9989	0.9754	0.9724	0.9578
15	0.9859	0.9946	0.9852	0.943	0.9899	0.9432	0.9781
16	0.9582	0.9488	0.9592	0.9601	0.9534	0.9553	0.9578
17	0.9178	0.9209	0.9167	0.9123	0.9189	0.9067	0.9167
18	0.8659	0.8534	0.8699	0.8612	0.8701	0.8523	0.8699
19	0.8043	0.7956	0.8078	0.7599	0.7943	0.8145	0.7659
20	**0.7348**	**0.3715**	**0.3821**	**0.3698**	**0.3765**	**0.3567**	**0.3288**
21	0.6596	0.6657	0.6566	0.6578	0.6602	0.5992	0.6603
22	0.5808	0.5743	0.5810	0.5899	0.5845	0.5956	0.5790
23	0.5007	0.5087	0.4998	0.5004	0.5078	0.4934	0.4967
24	0.4215	0.4187	0.4223	0.4279	0.4279	0.4278	0.4399
25	0.3451	0.3512	0.3489	0.3467	0.3398	0.3301	0.4307
26	0.6859	0.6904	0.6830	0.6823	0.6811	0.6745	0.6877

Both PSO and BFO are similar in approach and the development of these algorithms has been influenced by the behaviour of groups of living organisms. As a result, the difference between particle swarm optimization and bacterial foraging optimization is best brought out by studying their relative performance, i.e., the total computational time as shown in Tables 7.10–7.13. The comparison is performed for random configurations of one, two and three complete faults, and for a combination of complete and partial faults. Despite the fact that the both these techniques produced the similar results, it is clearly seen that PSO is faster than BFO for the above mentioned test conditions. This comparison is in agreement with results obtained by other researchers in the field of antenna engineering [Datta, and Misra, 2009]. Attempts are being made in order to develop faster versions of BFO [Datta *et al.*, 2008; Panikhom *et al.*, 2010]. However, some instances where BFO out-performs PSO have been reported [Biswas *et al.*, 2007].

Table 7.10 Time analysis for computation of one defective element (5 random configurations)

Failed Element Position	Time(s) (18 Samples)				Time(s) (35 Samples)			
	PSO		BFO		PSO		BFO	
1	1.52		106.89		2.62		208.34	
3	1.49	Average Time(s)	112.78	Average Time(s)	2.54	Average Time(s)	196.76	Average Time(s)
13	1.59		108.54		2.68		189.04	
18	1.46	**1.514**	101.34	**107.022**	2.73	**2.632**	210.73	**201.136**
25	1.52		105.56		2.59		200.81	

Table 7.11 Time analysis for computation of two defective elements

Failed Element Position	Time(s)(18 Samples)				Time(s) (35 Samples)			
	PSO		BFO		PSO		BFO	
6, 12	2.1		132.78		4.54		232.51	
9, 17	2.21	Average Time(s) **2.164**	119.67	Average Time(s) **131.616**	4.68	Average Time(s) **4.486**	221.84	Average Time(s) **235.108**
23, 25	2.12		147.09		4.32		256.90	
1, 13	2.16		132.67		4.39		234.71	
5, 21	2.23		125.87		4.50		229.52	

Table 7.12 Time analysis for computation of three defective elements

Failed Position	Time(s)(18 Samples)				Time(s) (35 Samples)			
	PSO		BFO		PSO		BFO	
3, 10, 16	2.56		167.09		4.45		223.65	
1, 17, 21	2.67	Average Time(s) **2.722**	140.87	Average Time(s) **156.04**	4.91	Average Time(s) **5.068**	267.94	Average Time(s) **250.16**
6, 12, 19	3.17		153.92		5.87		234.76	
11, 14, 23	2.78		151.48		4.14		275.49	
4, 15, 25	2.43		157.93		5.97		248.96	

Table 7.13 Time analysis for computation of combination of complete and partial defective elements

Failed Position	Time(s)(18 Samples)				Time(s) (35 Samples)			
	PSO		BFO		PSO		BFO	
4 and 16(100%), 20(50%)	3.78		120.95		4.67		240.72	
6 and 23(50%), 15(100%)	2.34	Average Time(s) **3.09**	167.96	Average Time(s) **149.308**	5.12	Average Time(s) **5.018**	300.87	Average Time(s) **268.59**
10 and 25(100%), 1 (50%)	4.01		145.98		4.96		276.83	
7 and 15 (100%), 24(50%)	3.33		158.39		5.33		256.66	
2 and 19(50%), 17(100%)	2.01		153.26		5.01		267.89	

7.5 Hybrid Technique

Section 7.2 of this chapter discusses the detection of faults in a linear array using the concept of neural networks. In this section, the problem statement is extended to the detection of faults in an 8 × 8 planar array comprising of the linear array designed on Section 7.2. The structure of the array is shown in Fig. 7.23. The array has a resonant frequency of 1.9 GHz. When free of faults, the array possesses a directional pattern as shown in Fig. 7.24.

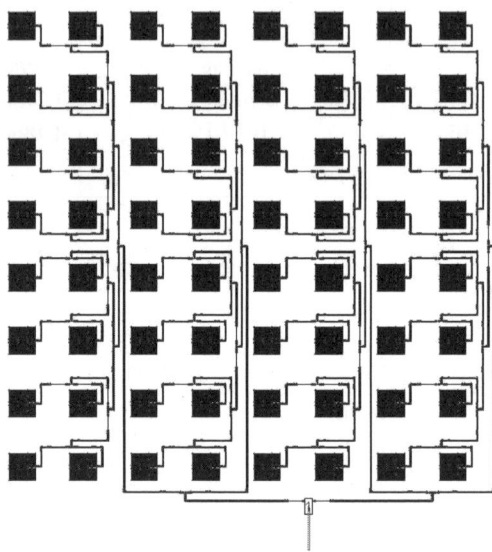

Fig 7.23 Schematic diagram of the 8 × 8 planar microstrip array

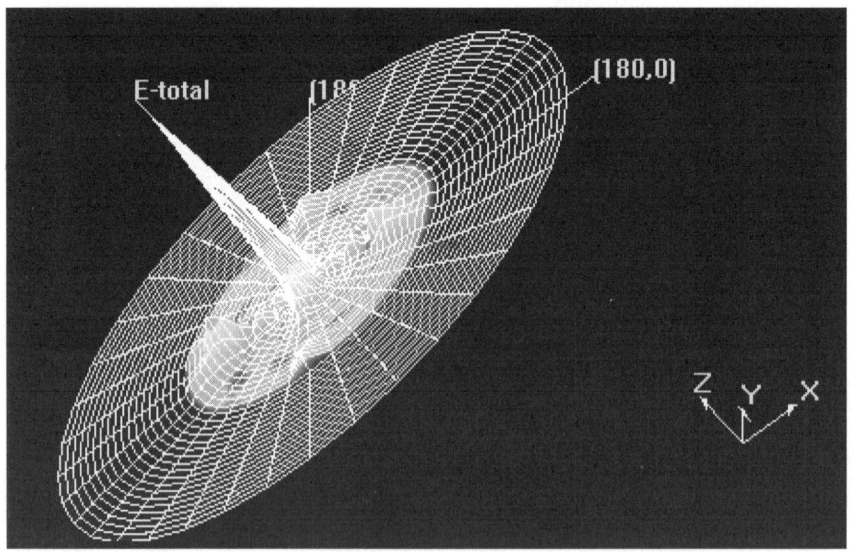

Fig 7.24 3D radiation pattern of the 8 × 8 planar microstrip patch array without any faults

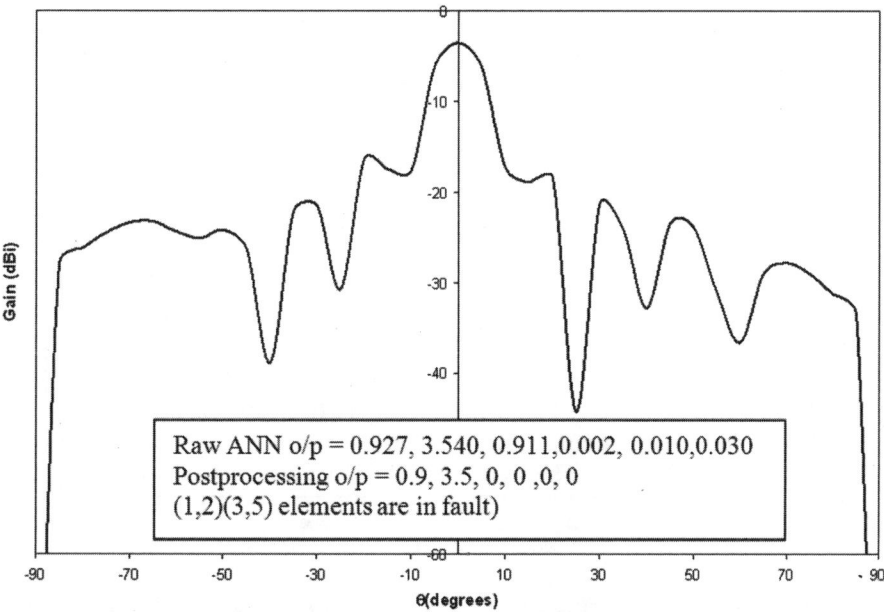

Raw ANN o/p = 0.927, 3.540, 0.911,0.002, 0.010,0.030
Postprocessing o/p = 0.9, 3.5, 0, 0 ,0, 0
(1,2)(3,5) elements are in fault)

Fig 7.25 Pattern for double (9th and 35th with coordinates (1,2), (3,5)) element fault in the planar array with the corresponding NN in the inset

Table 7.14 Network / Training parameters of the NN developed for fault finding in linear array

Parameters	Value
Neurons in the input layer	37
Neurons in the output layer	6
Number of hidden layers	1
Neurons in the hidden layer	27
Learning rate (η)	0.5
Performance ratio (γ)	0.7
Training tolerance	1×10^{-2}
Training time (on a 2.8 GHz dual core workstation, 4GB RAM)	12 minutes

The technique followed in this example is quite similar to the one followed in the example of the linear array. The total number of the faults in the array is restricted to six. This value corresponds to roughly 10% of the total number of elements in the planar array. The neural network parameters used in this simulation are given in Table 7.14. Random faults are generated and used for the training of the neural network. Some of these cases are shown in Figs. 7.25 and 7.26.

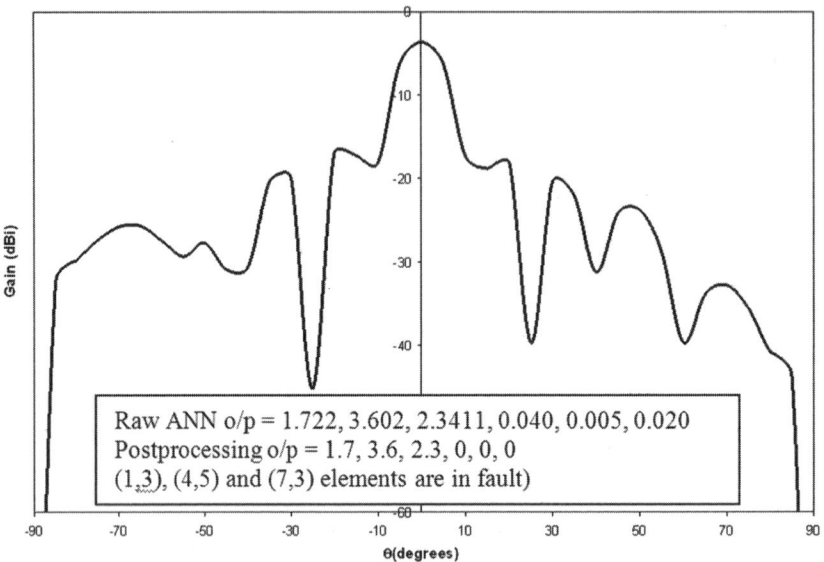

Raw ANN o/p = 1.722, 3.602, 2.3411, 0.040, 0.005, 0.020
Postprocessing o/p = 1.7, 3.6, 2.3, 0, 0, 0
(1,3), (4,5) and (7,3) elements are in fault)

Fig 7.26 Pattern for triple (17th, 36th, 23rd with coordinates (1,3), (4,5), and (7,3)) element fault in the planar array with the corresponding NN output in the inset

The overall performance of this neural network technique is improved by incorporating genetic algorithm (GA) in the computational process. Specifically, the problem is tackled using GA assisted backpropagation [Rajasekharan and Pai, 2003], in which the optimized weights of the neural network are obtained using GA [Choudhury *et al.*, 2009]. Results show that this hybrid algorithm decreased the error tolerance and at the same time produced the same results as those obtained using the neural network alone. This may be attributed to the fact that once a neural network is trained, it no longer depends on the training algorithm. In the classical backpropagation method, the training data is pre-processed and raw data from network is post processed in order to obtain the end result.

7.6   Summary

Detection of faulty elements in antenna arrays is one of the biggest operational challenges faced by antenna engineers. These faults degrade the radiation pattern of the antenna array. Commonly used techniques for fault detection involve the study of the near-field radiation pattern. This technique is not feasible in practical application where the antenna array is deployed on remote platforms such as spacecraft, etc.

In this chapter, a method of detecting these faults from the far-field information of the array using various soft computing techniques has been presented. These techniques include *neural networks* (NN), *particle swarm optimization* (PSO, and *bacterial foraging optimization* (BFO). Through simulation, all the three techniques have been found to be extremely effective in detecting the position and nature of the faults. The computation time, merits and shortcomings of these techniques have been presented. In the last section of this chapter, a hybrid technique, viz. genetic algorithm assisted backpropagation algorithm has been presented in order to decrease the computational time of the neural network based approach.

Once the faults are detected, suitable compensation techniques may be incorporated to restore the radiation pattern of the antenna array [Mailloux, 1996; Steyskal and Mailloux, 1998; Levitas *et al.*, 1999; Yeo *et al.*, 1999]. This increases the life-span of the array and at the same time saves the cost of replacement.

Conventional NN, PSO and BFO have been implemented in this chapter. While these methods have been proven effective from the results point of view, they are not efficient from the view of computational cost. Therefore, exploring the implementation of faster versions of these algorithms is desirable. Further, efforts could be made to integrate suitable pattern recovery methods with the fault detection algorithms. Such systems should be capable of detecting faults and providing compensation techniques for the same, thereby making it useful for practical applications.

References

Acharya, O. P., A. Patnaik, and S. N. Sinha, "Null steering in failed antenna array," *Applied Computational Intelligence and Soft Computing* (Hindawi), vol. 2011, Article ID 692197, doi: 10.1155/2011/692197., 2011.

Acharya, O. P., A. Patnaik, and S. N. Sinha, "Limits of compensation in a failed antenna array," *International Journal of RF and Microwave Computer-Aided Engineering*, vol. 24, no. 6, pp. 635–645, Nov. 2014.

Anderson, J. A. and E. Rosnefeld (eds.), *Neurocomputing: Foundation of Research*, Cambridge, MA, MIT Press, ISBN: 9780262510486, 752p., 1988.

Balanis, C. A., *Antenna Theory: Analysis and Design*, John Wiley and Sons, Inc., ISBN: 9780471714613, 1136p., 2005.

Biswas, A., S. Dasgupta, S. Das, and A. Abraham, "Synergy of PSO and bacterial foraging optimization – A comparative study on numerical benchmarks," *Innovations in Hybrid Intelligent Systems* (Springer-Verlag), vol. ASC 44, pp. 255–263, 2007.

Bregains, J. C., F. Ares, and E. Moreno, "Matrix pseudo-inversion technique for diagnostics of planar arrays," *Electronics Letters*, vol. 41, no. 1, pp. 7–8, 6th Jan. 2005.

Bucci, O. M., A. Capozzoli, and G. D'Elia, "Diagnosis of array faults from far-field amplitude-only data," *IEEE Transactions on Antennas and Propagation*, vol. 48, no. 5, pp. 647–652, 2000.

Chakrabarty, A., B. N. Das, and A. Bhattacharya, "Detection of localized array fault from near field data," *Antenna and Propagation Society International Symposium Digest*, vol. 3, pp. 1408–1411, June 1991.

Choudhury, B., O. P. Acharya, and A. Patnaik, "A PSO application for locating defective elements in antenna arrays," *World Congress on Natural and Biologically Inspired Computing*, pp. 1094–1098, Bhubaneswar, 2009.

Choudhury, B., O. P. Acharya, and A. Patnaik, "Bacteria foraging optimization in antenna engineering: An application to array fault finding," *International Journal of RF and Microwave Computer-Aided Engineering*, vol. 23, no. 2, pp. 141–148, Mar. 2013.

Choudhury, B., O. P. Acharya, and A. Patnaik, "Fault finding in antenna arrays using bacteria foraging optimization technique," *National Conference on Communications* (NCC 2011), pp. 1–5, Banglore, 2011.

Christodoulou, C. G. and M. Georgiopoulous, *Application of Neural Networks in Electromagnetics*, Nowrood, MA, Artech House, ISBN: 9780890068809, 512p., 2000.

Datta, T. and I. S. Misra, "A comparative study of optimization techniques in adaptive antenna array processing: The bacteria foraging algorithm and particle swarm optimization," *IEEE Antennas and Propagation Magazine*, vol. 51, no. 6, pp. 69–79, Dec. 2009.

Datta, T., I. S. Misra, B. B. Mangaraj, and S. K. Imtiaj, "Improved adaptive array for faster convergence," *Progress in Electromagnetic Research C*, vol. 1, pp. 143–157, 2008.

Gattoufi, L., D. Picard, A. Rekiouak, and J. Ch. Bolomey, "Matrix method for near-field diagnostic techniques of phased arrays," *IEEE International Symposium on Phased Array Systems and Technology Digest*, pp. 52–57, 1996.

Gattoufi, L., D. Picard, Y. Rahmat Samii, and J. Ch. Bolomey, "Regularized matrix method for near-field diagnostic techniques of phased array antennas," *IEEE Antennas and Prop. Soc. Int. Symp. Digest*, vol. 2, pp. 1066–1069, 1997.

Haykins, S., *Neural Networks: A Comprehensive Foundation*, New York: IEEE Press/ IEEE Computer Society Press, ISBN: 9780139083853, 842p., 1994.

Holland, J. H., *Adaptation in Natural and Artificial Systems: An Introductory Analysis with Applications to Biology, Control, and Artificial Intelligence*, University of Michigan Press, Ann Harbor, 1975.

Hornik, K., M. Stichcombe, and H. White, "Multilayer feedforward networks are universal approximators," *Neural Networks*, vol. 2, pp. 359–366, 1989.

Jang, J. S. R., C. T. Sun, and E. Mizutani, *Neuro-Fuzzy and Soft Computing: A computational Approach to Learning and Machine Intelligence*, Prentice Hall, NJ, ISBN: 9780132610667, 614p., 1997.

Kennedy, J., and R. Eberhart, "Particle swarm optimization," *Proceedings of IEEE International Conference on Neural Networks*, pp. 1942–1948, 1995.

Lee, J. J., E. M. Ferren, D. Pat Woollen, and K. M. Lee, "Near-field probe used as a diagnostic tool to locate defective elements in an antenna array," *IEEE Transactions on Antennas and Propagation*, vol. 36, no. 6, pp. 884–890, June 1988.

Levitas, M., D. A. Horton, and T.C. Cheston, "Practical failure compensation in active phased arrays," *IEEE Transactions on Antennas and Propagation*, vol. 47, no. 3, pp. 524–535, 1999.

Lord, J. A., G. G. Cook, and A. P. Anderson, "Reconstruction of the excitation of an array antenna from measured near-field intensity using phase retrieval," in *Proceedings of Institute of Electrical Engineers*, vol. 139, pp. 392–396. 1992.

Mailloux, R. J., "Array failure correction with a digitally beamformed array," *IEEE Transactions on Antennas and Propagation*, vol. 44, pp. 1542–1550, 1996.

Mailloux, R. J., *Phased Array Antennas Handbook*, Norwood, MA: Artech House, ISBN: 9781580536899, 496p., 1994.

Migliore, M. D. and G. Panariello, "A comparison of interferometric methods applied to array diagnosis from near-field data," *IEE Proceedings – Microwave Antennas Propagation*, vol. 148, no. 4, pp. 261–267, Aug. 2001.

Mishra, R. K., "An overview of neural network methods in computational electromagnetics," *International Journal of RF and Microwave Computer Aided Engineering*, vol. 12, no. 1, pp. 98–108, 2002.

Mishra, R. K. and A. Patnaik "Designing rectangular patch antenna using the neurospectral method," *IEEE Transactions on Antennas & Propagation*, vol. 51, no. 8, pp., 1914–1921, Aug. 2003.

Panikhom, S., N. Sarasiri, and S. Sujitjorn, "Hybrid bacteria foraging and tabu search optimization (BTSO) algorithms for Lyapunov's stability analysis of non-linear systems," *International Journal of Mathematics and Computers in Simulation*, vol. 4, issue 3, pp. 81–89, 2010.

Passino, K. M., "Biomimicry of bacteria foraging for distributed optimization and control," *IEEE Control Systems Magazine*, vol. 22, pp. 52–67, 2002.

Patnaik, A., and R. K. Mishra, "ANN techniques in microwave engineering," *IEEE Microwave Magazine*, vol. 1, no. 1, pp. 55–60, 2000.

Patnaik, A., B. Choudhury, P. Pradhan, R. K. Mishra, C. Christodoulou, "An ANN application for fault finding in antenna arrays," *IEEE Transactions on Antennas and Propagation*, vol. 55, no. 3, pp. 775–777, Mar. 2007.

Patnaik, A., D. Anagnostou, C. G. Christodoulou, and J. C. Lyke, "Neurocomputational analysis of a multiband reconfigurable planar antenna," *IEEE Transactions on Antennas and Propagation*, vol. 53, no. 11, pp. 3453–3458, Nov. 2005.

Peters, T. J., "A conjugate gradient-based algorithm to minimize the sidelobe level of planar arrays with element failures," *IEEE Transactions on Antennas and Propagation*, vol. 39, pp.1497–1503, 1991.

Proakis, J. G., and M. Salehi, *Digital Communications*, McGraw-Hill International Ed., ISBN: 9780071263788, 1150p., 2008.

Robinson, J. and Y. Rahmat-Samii, "Particle swarm optimization in electromagnetics," *IEEE Transactions on Antennas and Propagation*, vol. 52, pp. 397–407, 2004.

Rodriguez, J. A. and F. Ares, "Optimization of the perfromance of arrays with failed elements using simulated annealing technique," *Journal of Electromagnetics Wave and Applications*, vol. 12, pp.1625–1638, 1998.

Rodriguez, J. A., F. Ares, and E. Moreno, "GA procedure for linear array failure correction", *Electronics Letters*, 36, pp. 196–198, 2000.

Steyskal, H. and R. J. Mailloux, "Generalization of a phased array error correction method," *IEEE Antennas and Propagation Society, AP-S International Symposium (Digest) Proceedings of the 1996 AP-S International Symposium & URSI Radio Science Meeting.* pp. 506–509, Jul. 1996.

Su, C. and S. M. Lin, "A method for locating defective elements in the large planar array," *IEEE Antenna and Propagation Society International Symposium Digest*, vol. 24, pp. 31–33, 1986.

Wang, L. L., D.G. Fang, and W. X. Sheng, "Combination of genetic algorithm and fast Fourier transform for synthesis of arrays," *Microwave and Optical Technology Letters*, vol. 37, pp. 56–59, 2003.

Yeo B. K. and Y. Lu, "Adaptive array digital beam forming using complex-coded particle swarm optimization-genetic algorithm," *Proceedings of Asia-Pacific Microwave Conference*, 3p., 4–7 Dec. 2005.

Yeo, B. K. and Y. Lu, "Array failure correction with a genetic algorithm," *IEEE Transactions on Antennas and Propagation,* vol. 47, pp. 823–828, 1999.

Zadeh, L. A., *Fuzzy Logic, Neural Networks and Soft Computing.* One-page course announcement of CS 294-4, Spring 1993, The University of California at Berkeley, Nov. 1992.

Zainud-Deen S.H., et. al., "Array failure correction with orthogonal methods," *Proceedings of the 21st National Radio Science Conference,(NRSC 2004)*, pp. B7:1–9, March 2004.

Zhang, Q. J. and K. C. Gupta, *Neural networks for RF and Microwave Design*, Norwood, MA, Artech House, ISBN: 9781580531009, 369p., 2000.

Multi-Objective Particle Swarm Optimization for Active Terahertz Devices

Arya Menon*

The terahertz spectrum is gaining momentum in a wide range of fields including astronomy, material characterization, tumour detection, security and detection of concealed items, etc. among others. The main obstacle in the rapid growth of terahertz applications is the lack of natural dielectrics, and of high power, efficient sources. Emergence of metamaterial science and technology has attracted researchers to design and develop terahertz devices using these artificially engineered materials. These metamaterial structures are highly resonant, restricting the terahertz applications to narrow bands. This issue can be sorted out by designing active metamaterials.

As discussed in the Chapter 5, mathematical formulation for the design of metamaterial based structures is complicated and soft computing has proven to be an efficient method to arrive at design solutions. Further, terahertz devices themselves often are complex designs being multilayer in nature with an embedded metamaterial layer. Further, to make the devices tunable, mechanisms such as the MEMS switches, diodes, etc., should be incorporated in the design. In order to achieve an optimized design with these complex mechanisms one ought to go for a multi-objective optimization computational engine. This issue has been discussed in detail in this chapter via a tunable terahertz absorber design.

8.1 Introduction to Terahertz Technology

Terahertz refers to the range of frequencies that lies within the microwave and infrared (IR) bands. It can be loosely defined as the band between 0.1–10 THz. At times, the term

---

* This chapter adapts Ms Arya Menon's dissertation entitled: *Active Terahertz Metamaterials for Biomedical Applications*, submitted to Manipal University in May 2014. This work was carried out at CSIR-National Aerospace Laboratories, Bangalore, India.

"submillimetre band" is used to describe frequencies lying in the 0.3–3 THz band. While systems and applications operating in the microwave and IR ranges have been well *established* over decades, the development of terahertz technology has been slow. However, in recent times, this technology has evolved rapidly to find major applications relating directly to human lives, especially in the field of biomedical imaging and sample identification.

8.1.1 Properties of terahertz spectrum

The primary reason for this slow developmental trend was the lack of terahertz sources during the early years, coupled with the fact that this range displays a natural break-down point in electric and magnetic properties in conventional materials [Ferguson and Zhang, 2009, Smith *et al.*, 2004]. This property put constraints on the material used in the fabrication of terahertz devices. Other properties of terahertz radiation include straight line propagation, non-ionizing nature, and ability to penetrate a wide variety of non-conducting materials (however, the penetration depth is lower than of microwave).·

It is seen that the absorption coefficient of terahertz radiation in water ranges between 100 and 1000 cm^{-1} [Seigel, 2004]. Therefore, terahertz radiation suffers from very high attenuation during propagation in the atmosphere. In an article titled "*The Truth about Terahertz*" published in IEEE Spectrum in August 2012, Carter Armstrong examined the possible applications, areas of future work and limitations of terahertz technology. He observed that a 1 THz signal transmitted at 1 W retains only 10^{-13} percent of its signal at a distance of 1 km. This implies the need for very high transmission power. Hence, terahertz radiation is not used for long distance communication or radar applications.

8.1.2 Applications

Recent advances in the field of terahertz sources and novel terahertz materials have brought about potential applications in earth-based systems that take advantage of the otherwise cons of terahertz [Ferguson and Zhang, 2009]. These applications are primarily based on molecular sensing and can be used in space, environmental, biotechnological, security, and biomedical applications [Maagt, 2007]. Further, imaging in the terahertz offers higher resolution compared to microwave frequencies as the diffraction limit is given by 1.22λ. However, while IR, optical, and MRI systems are known to provide resolution greater than terahertz systems, the latter offer better contrast and hence are more useful for distinguishing substances [Seigel, 2004].

8.1.2.1 Space platform

Historically, terahertz technology has been employed in space-based applications such as earth observation science, radio astronomy, and planetary/cometary science. In fact, terahertz spectroscopy offers an insight into the physical condition (heat, pressure, etc.) as well as molecular concentration of the region being observed. This property is attributed to the fact that terahertz photons are emitted during changes in a molecular, rotational or bending state of atoms [Maagt, 2007]. Terahertz systems have already been deployed in missions such as the NASA Cosmic Background Explorer (COBE), Microwave Instrument for the Rosetta Orbiter (MIRO), etc.

8.1.2.2 Security

Terahertz radiation is capable of detecting molecular signatures of materials. This interesting property is being exploited for developing systems for security, especially airport security systems. These systems have the capability to detect unwanted hidden objects like knives, guns, etc. Additionally, terahertz molecular signatures also enable the detection of illegal and harmful liquids and drugs [Maagt, 2007]. For security applications, incoherent terahertz radiation is often used.

8.1.2.3 Biomedical field

The non-ionizing nature of terahertz radiation offers great potential in the medical field and could be used to replace contemporary scanning technology with harmless terahertz systems. Mark Stringer of Leeds University reported on studies by various research groups on the effects of terahertz radiation on human tissue; exposure of human keratinocytes up to 10 µW of radiation and lymphocytes up to 1 mW of radiation did not reveal any detrimental result. The reflection coefficient for tissue samples have been reported to vary with water content [Seigel, 2004]. Hence, when terahertz radiation is incident on tissue samples, penetration depth and reflection varies depending on the water content in the tissue. This offers a method to develop high contrast images based on water content and has been used in the examination of severity of burns and tumour morphology.

Further, terahertz time domain spectroscopy (THz–TDS) enables measurements of states of proteins and oligonucleotides, avidin–biotin binding and DNA hybridization using molecular signatures. Research is also being carried out to detect unwanted polymorphs in prepared drugs using THz–TDS. Enhanced contrast offered by terahertz also opens up avenues for easy detection of tumours and cancerous tissues. The ·operating principle behind this application is based strongly on the fact that cancerous tissues are observed to contain more blood vessels than normal tissue [Kearney, 2013]. Therefore, the water content in these tissues is higher than that of the surrounding tissues. Differential reflection of terahertz radiation can hence be used to identify them.

8.1.3 Challenges of terahertz technology

As mentioned previously, development in terahertz imaging has not been on par with its neighbours in the frequency spectrum (microwave and IR). This slow trend is due to the multiple challenges in terahertz design, which are mentioned below.

8.1.3.1 Material issues

As mentioned previously, the dearth of naturally available materials that show desirable properties at these frequencies is one of the biggest impediments in the development of devices. In order to fully understand why such a characteristic should occur, it is imperative to study the interaction of molecules with the incident terahertz wave.

Electric fields have the ability to polarize dielectrics. When a dielectric is exposed to an alternating field, the direction of polarization varies with the field. However, the response of the material is not instantaneous with respect to the field and the switching of the direction of polarization occurs after a certain amount of time, called the *relaxation time*. As one enters the terahertz region, the time period of the field becomes smaller than the relaxation time; the polarization of the material ceases to keep up with the incident field.

In practice, three different polarization mechanisms operate in dielectrics: electronic mechanism, ionic mechanism and orientational mechanism. Each of these mechanisms is characterized by a resonant frequency. In the terahertz region, as the frequency increases, each mechanism ceases to exist. Further, it should be noted that for small frequency increments beyond each resonant frequency, the polarization response drops significantly. This results in a sharp dip in the dielectric constant of the material immediately after each resonance and the subsequent switching off of each mechanism. Ultimately, at around 1000 THz, all mechanisms stop operation and the dielectric constant of the material becomes 1.

This issue can be resolved by the usage of certain artificially engineered materials called metamaterials. The designer is capable of engineering the dielectric constant of metamaterials to suit the design requirement. These metamaterials also show some interesting properties and find application in many electromagnetic applications today, including in the design and fabrication of terahertz devices.

8.1.3.2 Design issues

As mentioned previously, metamaterials are used for the development of terahertz devices. These metamaterials are subwavelength structures, the analysis of which is carried out using advanced electromagnetic concepts like equivalent circuit analysis (ECA), FDTD method, etc. However, mathematical relationships that can predict the resonant frequency of the metamaterial as functions of its dimensions are not currently available in literature. As a result, metamaterial design is a cumbersome and often iterative task.

8.1.3.3 Fabrication issues

Two main fabrication issues must be considered for the design of terahertz devices: (a) identification of suitable fabrication technique for the realization of metamaterial designs in the terahertz range, and (b) modification of fabrication techniques in order to enable cheap, mass production.

8.1.3.4 Characterization issues

Accurate characterization of terahertz devices can be performed when the designer has access to high performance terahertz sources and detectors. Currently, development of gyrotrons, backward wave diodes, and resonant tunnelling diodes has enabled generation of 0.3–1 THz. The output powers of these devices are in the range of μW–mW. Research is currently being conducted in order to develop high power, efficient sources. In addition, the development of detectors with high sensitivity and low loss is required in order to detect this low-powered radiation.

8.2   Trends in Active Terahertz Devices

Metamaterials are EM patterns of metallic strips over dielectric substrates and are narrowband in nature. However, for application in the biomedical field, space, etc., the designed metamaterial must be resonant over a wide range of frequencies. Many techniques have been proposed in order to change the resonant frequency of metamaterials. These techniques include MEMS based tuning, photo excitation, electrical actuation, and thermal actuation.

These tuning mechanisms have been implemented for metamaterials operating in the microwave and IR frequency bands. Ekmekei et al. [Ekmekei et al., 2009] proposed tuning of an X-band metamaterial using micro-split rings. The tuning is achieved by shorting multiple micro splits in the structure. It was observed that the presence of micro splits increased the resonant frequency due to the increase in the series capacitance. A method to control the micro splits using MEMS switches was also proposed.

Shadrivov et al. [Shadrivov et al., 2008] developed a nonlinear, negative refractive indexed metamaterial by loading a variable capacitor on the gap of a split ring resonator. This was realized using a varactor diode and therefore, tuning was achieved by varying the bias voltage to the diode. Hence, both the positive and negative refractive indices can be obtained depending upon the bias. Further, by using zero bias diodes, switching between high and low reflection states can be achieved. It was possible to obtain a tuning range of 0.5 GHz using this structure.

Prathiba et al. [Prathiba et al., 2009] dispersed gold nano-particles (GNP) in a liquid crystal (LC) medium. This enabled the creation of a metamaterial that can be tuned by changing the molecular orientation of the liquid crystal. In addition, the optical properties can be tuned by changing the volume of GNP. The advantages of such a metamaterial with respect to traditional lithographic techniques such as, creation of bulk metamaterials and cheaper fabrication technique is presented. In this section, a review on the application of the above mentioned tuning mechanisms for terahertz metamaterial design is discussed.

8.2.1 MEMS based tuning

Tao et al. [Tao et al., 2008] designed and fabricated a novel balanced SRR structure with a bi-material cantilever beam mounted on its gap. The balanced SRR structure couples negligibly to a magnetic field where as strongly to an electric field, unlike a regular SRR, which couples strongly to both. The cantilever, in the initial state, is bent due to micro-machining residual stress and therefore, shorts the SRR. On increasing the temperature, due to the differential expansion in aluminium and silicon nitride—the two materials in the beam—the beam is forced to bend upwards. The surface profile of this structure was studied and results showed a transmission resonance of 0.75 THz. The structure has potential application in THz switches and modulators.

Ozbey et al. [Ozbey et al., 2011] designed a modified split ring resonator with a single end fixed cantilever incorporated into it. The cantilever can be dynamically controlled using an external magnetic field. This metamaterial presents the capability to be tuned over a continuous range of frequencies over wide bands rather than discreet frequencies or specific narrow bands. Two different designs were simulated in the tuning frequency range of 0.147–0.275 THz and 0.265–0.514 THz. The factors affecting the performance of the cantilever beam was also discussed.

Zhu et al. [Zhu et al., 2012] designed, fabricated and studied the performance of a metamaterial with anisotropy that could be tuned from positive to negative values and can be operated in the range of 1–5 THz. The basic structure of the designed metamaterial is a Maltese cross pattern that is repeated periodically. Anisotropy is introduced by displacing one of its beams, which is patterned onto a micro-actuator.

Yang *et al.* [Yang *et al.*, 2013] designed and discussed the fabrication of a tunable metamaterial absorber employing MEMS based actuation. The basic structure of the metamaterial is formed by two back-to-back split ring resonators separated from a ground plane by a dielectric. One of the split rings is placed on a movable beam. Hence, the resonant frequency of the metamaterial is varied by changing the distance between the two rings. Different MEMS based actuation techniques are suggested for controlling the beam. On simulation, it was observed that as the distance between the two rings was varied from 0–5μm, the resonant frequency changed from 1.02 THz to 1.20 THz.

Vendik *et al.* [Vendik *et al.*, 2012] proposed the designs of two different metamaterial structures whose resonant frequency could be tuned using a MEMS cantilever. The first structure that was proposed was a U-shaped resonator consisting of three strips connected via cantilever beams, which has the potential to be used as a tunable band-stop filter. When the bending angle of the cantilever beam was changed from 0° to 15°, the resonance frequency was found to change from 0.384 THz to 0.586 THz. In addition, a metamaterial that consists of a metal plate–dielectric–metal plate structure was also proposed. One such metal plate was fabricated so as to behave as a cantilever beam. Hence, the distance between the two metal plates could be varied, which in turn varied the resonant frequency. The variation of the resonant frequency and transmission coefficient with bending angle of the cantilever was simulated.

Zhang *et al.* [Zhang *et al.*, 2013] designed and fabricated a metamaterial that can be tuned to change its property from isotropic to anisotropic. The structure consisted of rows of gammadion shaped structures. Alternate rows were released from the substrate and could be moved in the lateral direction with the help of electromechanical actuators. This enabled the change of the relative positions of the gammadion structures and hence switching between isotropy and anisotropy was possible. This structure enabled polarization-dependent control over the transmitted wave as well as control on the transmission intensity and phase.

8.2.2 Photo excitation

Padilla *et al.* [Padilla *et al.*, 2006] studied the response of SRR array embedded in a GaAs substrate, which worked as an absorptive metamaterial at 0.5 THz and 1.6 THz. The dynamic control of the metamaterial by impinging the substrate with light was proposed. The effect of the power of the incident light on different absorption frequencies was studied.

Kozlov *et al.* [Kozlov *et al.*, 2011] designed a metamaterial using a cubic dielectric resonator as the basic unit. A metallic strip with a gap was etched onto one side of this dielectric cube. Tuning was achieved by varying the electrical length of the metal strip by photo-excitation of the gap or by voltage biasing. The structure was observed to act as a band pass filter with a pass-band from 0.45 THz to 0.5 THz.

Chowdhury *et al.* [Chowdhury *et al.*, 2012] designed and tested a dynamically reconfigurable split-ring resonator from fundamental resonant frequencies of 0.6 THz and 1.76 THz that could perform mode renormalization from odd-order modes to even-order modes. The design involved a split ring resonator with a silicon island placed at the gap. The charge carrier concentration in the gap could be dynamically varied by changing photon pump power used for photo-excitation. At higher pump powers, the odd-modes switched off and even-mode resonance was observed.

8.2.3 Electrical actuation

Paul *et al.* [Paul *et al.*, 2009] designed a polarization independent terahertz modulator. The structure consisted of an array of gold cross on an n-doped GaAs substrate, which was externally biased. The crosses were interconnected using 1 μm wires. For THz frequencies, the structure was seen to be transparent during the forward bias and opaque during the reverse bias. Therefore, the structure could modulate THz waves if the bias was a modulating signal. The modulation speed was studied for modulating signals up to 100 kHz. A cut-off frequency of 80 kHz was observed which was much greater than that of semiconductor based modulators. Further, symmetric structure rendered the metamaterial polarization independent.

Kowerdziej *et al.* [Kowerdziej *et al.*, 2012] developed and analysed the performance of a metamaterial transducer that employs *in-plane switching mode* (IPS). The structure consisted of a nematic liquid crystal, sandwiched between two Ω shaped resonators. The orientation of the crystal could be controlled by providing an external bias voltage. This tuning enabled one to make the refractive index positive, negative or zero. Furthermore, the band width over which the crystal displays negative refractive index could be tuned.

8.2.4 Thermal actuation

Němec *et al.* [Němec *et al.*, 2009] designed and fabricated a temperature controlled terahertz metamaterial. The structure consisted of rows of high permittivity rods made of $SrTiO_3$ (STO) of a certain width and a constant gap. Frequency tuning was achieved by changing the permittivity of STO by varying the temperature. The transmission spectra was obtained numerically using multidimensional transfer matrix method and experimentally using time domain terahertz spectroscopy, and a comparison was made between the two. The range of frequencies that showed negative permeability was found to undergo a red-shift on decreasing the temperature. Further, it was observed that resonance was dependent only on the geometry of the rods and not on their coupling. This could lead to potential application of this structure in the creation of bulk metamaterials.

8.3   Design of Terahertz Device

Components, such as switches, modulators, filters, amplifiers, etc., are essential for the generation and reception of terahertz radiation in terahertz systems. These devices are capable of conditioning and /or controlling the radiation, in order to make it suitable for a desired application. This section explores the trends in the design and fabrication of such terahertz devices.

Lenses: Low spatial resolution is a desirable characteristic for any imaging system and can be achieved using lens with strong focusing capabilities. Therefore, a good terahertz imaging system requires the use of THz lenses. Neu *et al.* [Neu *et al.*, 2011] proposed a lens that could focus terahertz radiation to a spot one wavelength in diameter. This lens was a gradient index lens and was realized using a three-layer metamaterial structure.

Headland *et al.* [Headland *et al.*, 2013] designed a compact, lattice-based metamaterial lens capable of deflecting a terahertz beam. Beam deflection up to 45° was reported. The thickness of the lens was less than a millimetre.

Modulators: Terahertz radiation is used in a modulated form in many applications. This implies the need for terahertz modulators. Further, in order to be robust from the view of potential applications, it would be desirable to design a modulator that can work over a broad-range of frequencies. However, the inherent resonant characteristics of metamaterials ensure that modulators are confined to work with a fixed, narrow frequency band. Therefore, a need for tunable terahertz modulators arises.

Padilla *et al.* [Padilla *et al.*, 2009] designed a Schottky diode-based active terahertz metamaterial that demonstrated 50% modulation in the transmitted radiation intensity. This was achieved by using the Schottky diode to vary substrate properties. The same device also found application as a filter. The fabrication techniques required to realize this modulator were shown to be compatible with contemporary semi-conductor fabrication techniques. Modulators may also be designed using microstrip resonators [Sheng *et al.*, 2011a], by controlling resonance through *high electron mobility transistor* (HEMT) embedded in the gaps of split ring resonators [Rout *et al.*, 2010], etc.

Emitters and Detectors: Considering that all existing applications of terahertz involve imaging in one form or the other, terahertz emitters and detectors are two devices that find a place in all terahertz systems.

Alves *et al.* [Alves *et al.*, 2012] designed complex, multi-band metamaterial films for emission of terahertz radiation between 4 and 8 THz with a bandwidth of 1 THz. Kan *et al.* [Kan *et al.*, 2012] used Smith Purcell effect in order to generate terahertz radiation. It was seen that electron bunching using 2 mm metallic gratings produced radiation of frequencies up to 0.7 THz.

Detection of terahertz also requires detectors made of special materials like metamaterials. In some cases, IR detectors such as bolometers or micro-bolometers may be used after minor modifications to their structure. Strikwerda *et al.* [Strikwerda *et al.*, 2011] proposed a bi-material cantilever based metamaterial THz detector with an optical read-out. This detector can be used for sensing and detection. Further, Kearney *et al.* [Kearney *et al.*, 2013] proposed the usage of metamaterial absorbers for improving the performance of bolometers in the terahertz spectrum.

Filters: Filters are essential components in any system and help remove unwanted back ground radiation. For example, terahertz is emerging to be a viable spectrum for inter-satellite links [Hwu *et al.*, 2013]. Terahertz filters are required in order to filter out unwanted, cosmic, background terahertz radiation. Lu *et al.* [Lu *et al.*, 2010] realized a higher-order, broadband, multi-layer, metamaterial based filter that operated between 227 and 283 GHz. The filter showed two transmission peaks in the specified bandwidth. A multilayer metamaterial filter consisting of chiral elements was also proposed by Sabah *et al.* [Sabah *et al.*, 2011a].

Filters may also be realized using metamaterial resonators that control surface waves using novel metamaterial structures as shown by Chen *et al.* [Chen *et al.*, 2011] and Horestani *et al.* [Horestani *et al.*, 2013]. Such designs behave as band pass filters. Further, Basiry *et al.* [Basiry *et al.*, 2012] used the periodic method of moment method for the design of optimized terahertz filters. Such novel methods decrease the computational time required to obtain optimized metamaterial filter designs.

8.3.1 Design of terahertz absorber

Terahertz absorbers have the potential to improve the performance of terahertz detectors [Kearney, 2013]. In order to illustrate the effectiveness of the usage of soft computing for design of terahertz devices, the design and optimization of a terahertz absorber is discussed in this chapter. As seen before, terahertz devices are generally based on metamaterials due to the lack of availability of natural dielectrics. For this example, a circular split ring resonator (CSRR) is chosen as the metamaterial unit cell. The design of the terahertz absorber has a step-by-step process involving: (i) design of CSRR in GHz spectrum, (ii) scaling of obtained structure for operation in THz, and (iii) addition of ground plane, other substrates, etc.

The structural parameters of the split ring resonator are obtained using a PSO based in-house developed CAD package, which is presented later in Chapter 9.

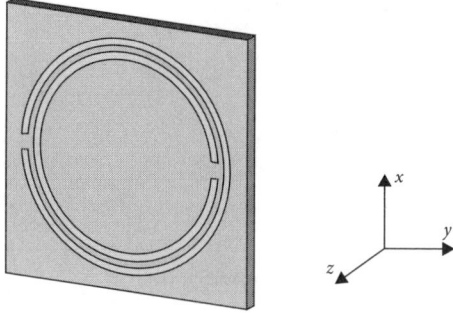

Fig 8.1   Circular split ring resonator for 2 GHz

Although, the algorithm of PSO is given in Chapter 2 of this book, the problem specific PSO parameters assigned for optimization of this CSRR design is given in Table 8.1. A polyimide substrate with dielectric constant of 3.5 is chosen for this design. Since the CAD package is developed for designs in GHz, an established technique, viz. *scaling* is used to obtain THz CSRR designs from GHz CSRR designs.

Table 8.1   PSO parameters used for design optimization of CSRR

Parameters	Value	Use
Wt	0.25	Inertial weight
C1	0.5	Constant1, determine relative pull of *pbest*
C2	0.5	Constant 2, determine relative pull of *gbest*
Np	10	Number of particles
Nd	4	Number of dimensions
Nt	30	Number of time steps

For resonance at 2 THz, the external radius of the outer ring is found to be 5.2 mm, the width of the rings are taken to be 0.3 mm, the distance between the two rings is 0.3 mm, the gap length in the two rings is taken as 0.75 mm, the substrate is of dimension 6 mm × 6 mm × 0.8 mm. This structure is simulated for free-space plane wave incidence in the y-direction in Fig. 8.1. Using parameter retrieval technique mentioned above, the permeability of the structure is obtained as shown in Fig. 8.2. As expected, the curve for permeability shows a Drude–Lorentz characteristic.

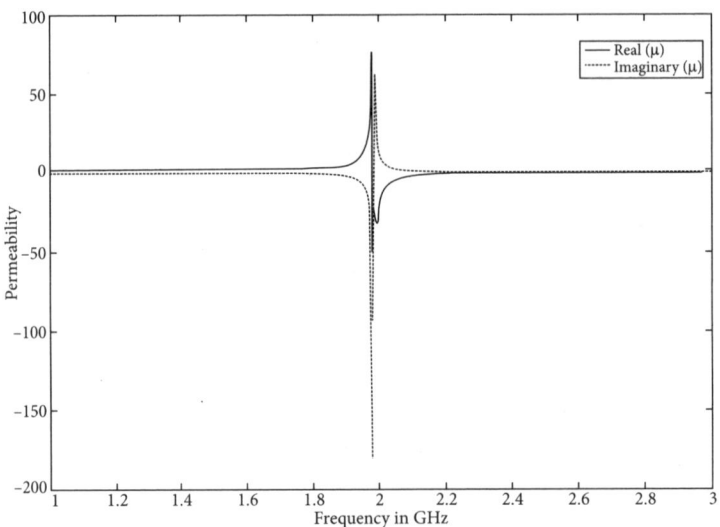

Fig 8.2 Permittivity of designed circular SRR for 2 GHz. The curve follows a Drude–Lorentz characteristic

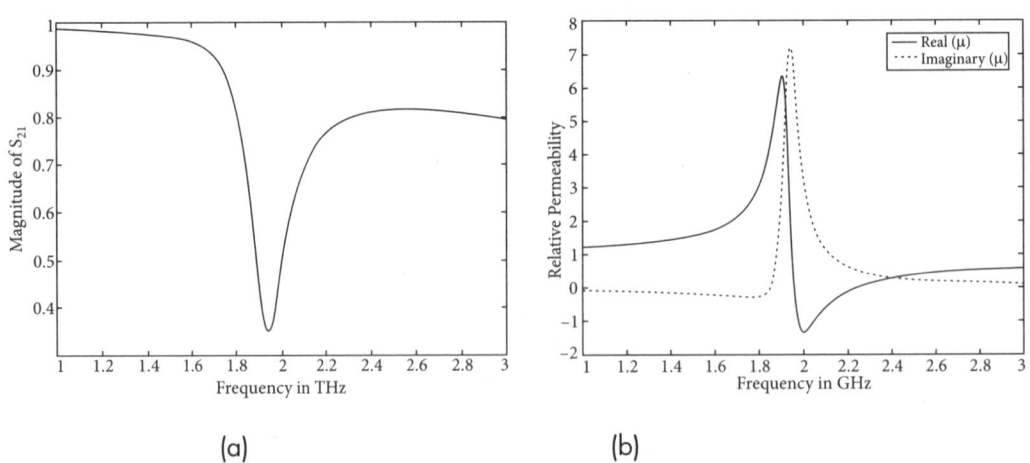

(a)   (b)

Fig 8.3 (a) S_{21} of circular SRR designed for 2 THz (b) Relative permeability of circular SRR designed for 2 THz

The designed structure is scaled and simulated. The transmission or scattering parameter S_{21} of the metamaterial structure is as shown in Fig. 8.3(a). Figure 8.3(b) shows the extracted relative permeability of the scaled circular split ring resonator.

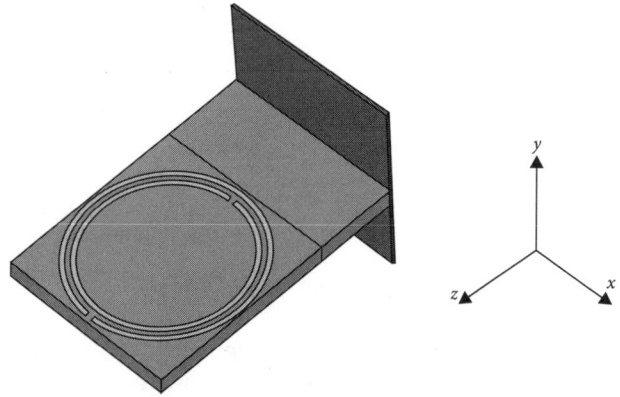

Fig 8.4 Optimized absorber design resonating at 2 THz

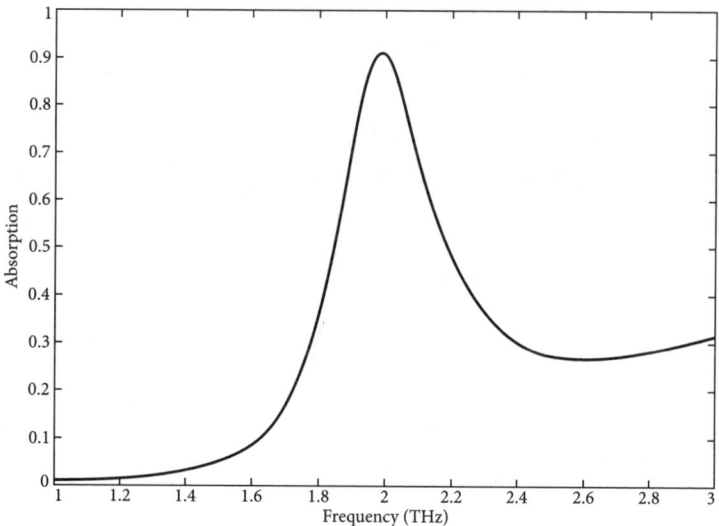

Fig 8.5 Absorption shown by absorber using designed unit cell (Fig. 8.4)

The absorber is then designed by stacking an additional substrate and a ground plane in order to make both reflection and transmission equal to zero (Fig. 8.4). The absorption is obtained using the following formula and plotted as shown in Fig. 8.5.

$$A = 1 - S_{11}^2 \tag{8.1}$$

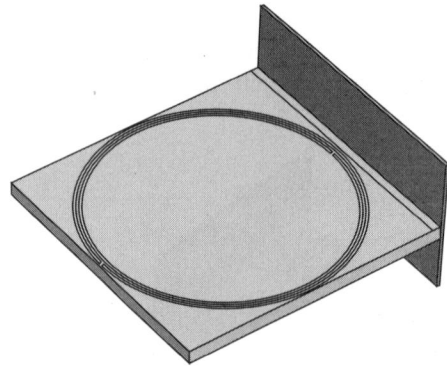

Fig 8.6 Schematic diagram of absorber for 1 THz

The same procedure was followed in order to design an absorber for resonance at 1 THz. The final structural parameters of the circular split ring resonator are as follows: Radius of outer ring = 7.3 μm; width of each ring = 0.1 μm; gap in each ring = 0.2 μm; and thickness of metal strip = 0.01 μm. This SRR is etched onto a polyimide substrate of dimensions 14.9 μm × 14.9 μm × 0.8 μm (Fig. 8.6).

8.3.2 Performance enhancement analysis

For the 2-THz absorber given in Fig. 8.4, it is observed that the peak performance just touched 90%, which is not enough for any application. One reason for this poor performance could be the fact that the dielectric material was not thick enough in the direction of wave incidence [Kearney, 2013]. Another possible reason for the poor performance could be large value of

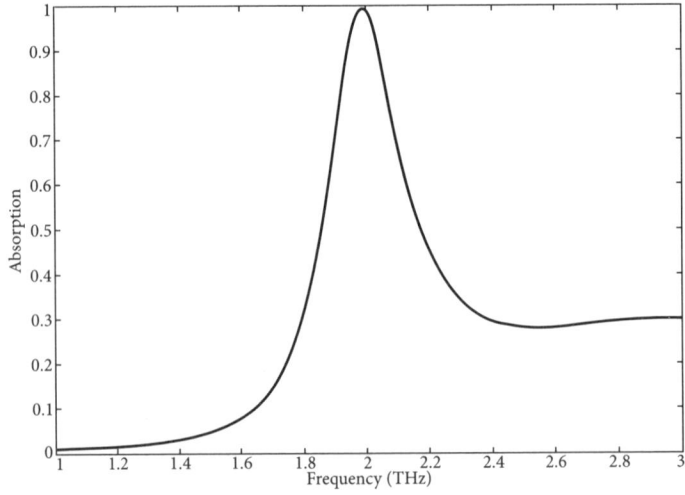

Fig 8.7 Absorption characteristics of 2 THz absorber

pitch or periodicity. The boundary conditions assigned in the simulations emulate periodic boundaries. In this case, the periodic boundaries are equal to the length and width of the substrate. It has been established that decrement in the pitch increases the absorption [Kearney, 2013].

The pitch of the unit-cell is reduced to 10.7 μm in the x-direction. The dielectric slab placed in between the split ring resonator and the ground plane was assigned a length of 4.7 μm. These values are obtained through manual iterations and the process is very time consuming and cumbersome.

The resonant frequency is found to shift when these changes in dimensions are incorporated. As a result, the structural parameters are iteratively changed in order to obtain resonance at 2 THz. For best absorption, the outer ring radius is 5.2 μm, the width of each ring is 0.3 μm, the distance between the two rings is 0.275 μm, the gap in each ring is 0.35 μm, height of substrate is 0.8 μm, thickness of metal is taken as 0.01 μm, and the ground plane for each unit cell is 10.7 μm × 8.2 μm × 0.1 μm. For support, a polyimide dielectric of the same dimensions is placed behind the ground plane.

The absorption characteristic of the structure in Fig. 8.4 is given in Fig. 8.7. The absorption was found to be 98.93% at 2 THz. It must be pointed out that the space above and below the circular SRR (and in front of the ground plane) is filled with air for simulation. This design cannot be used in practice. Hence, this space must be filled with a material that has the same permittivity as air. One example of such a material is Styrofoam.

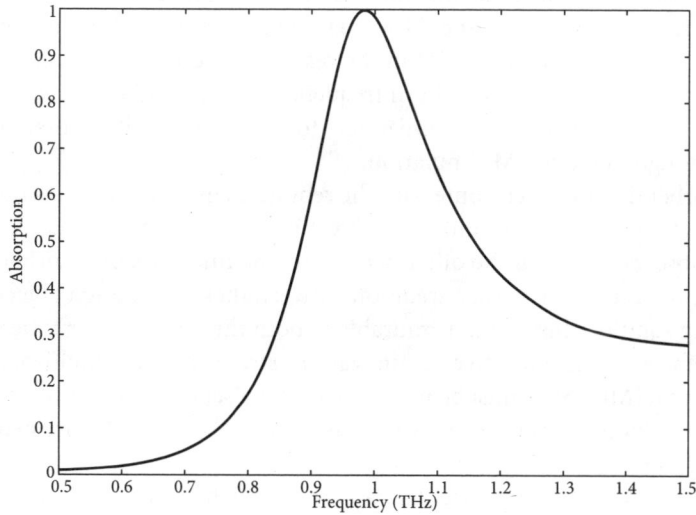

Fig 8.8   Absorption characteristic of 1 THz absorber

A similar procedure is followed in order to enhance the performance of the 1 THz absorber. An additional substrate of dimensions 14.9 μm× 0.5 μm × 0.8 μm is placed behind the SRR in the direction of wave propagation. This assembly is backed by a ground plane. The absorption characteristic of the structure is shown in Fig. 8.8. At 1 THz, the absorption is found to be 98.73%.

Therefore, it is seen that conventional absorber design procedures yield absorbers with approximately 98% absorption. These design procedures involve manipulation of the structural parameters of the absorber in order to produce accurate results. Therefore, there is scope for improving the performance for these absorbers in a time efficient manner.

8.4   Soft Computing for Performance Enhancement

From Chapter 4, it is seen that soft computing techniques, particularly, the *particle swarm optimization* (PSO) holds great potential for the design of absorbers. PSO is used to optimize a mathematical function that contains the information about the structural parameters of metamaterial absorber and their relationship with the overall absorption characteristic. However, it may not be possible to obtain such relationships when complicated geometries and novel material are used. Further, the optimization algorithm is single-objective oriented and this feature often leads to bulky absorber designs. In order to address these two scenarios, development of a *multi-objective* PSO (MOPSO) based computational engine is presented in this section, where an MOPSO kernel is integrated with a commercially available EM solver.

8.4.1 MOPSO based computational engine

While the theory behind the working of absorbers is well known, mathematical formulation for the design of metamaterial based absorbers is not easily available. Equations for the design of the metamaterial itself may be found. However, the presence of additional dielectrics in the construction of absorbers was found to shift the resonant frequency. The metamaterial design equations could not account for this shift in frequency. This lack of mathematical formulation necessitates the integration of the optimization tool with EM simulators so as to achieve optimization through iterative EM simulation.

Further, absorber designs often come with the requirement of being as less bulky as possible. The designer has to deal with two conflicting objective functions: on one hand, designs require extremely thin absorbers and on the other decreasing the thickness of absorbers decreases the absorption. One is forced to arrive at a trade-off between these two design objectives. Manually trying to obtain a solution that is best favourable to both these conditions is nearly impossible. Therefore, advanced multi-objective optimization strategies like multi-objective particle swarm optimization (MOPSO) must be used to arrive at a set of solutions wherein one value of fitness cannot be reduced without severely increasing the others. The theory behind MOPSO is discussed in Chapter 2.

In this chapter, a MOPSO code acts like a kernel, which calls a finite integration method (FIT) based EM solver so as to determine the fitness as shown in Fig. 8.9. This technique involves calling solver-specific commands (such as visual basic language (VBA), etc.), in the Matlab framework. The MOPSO algorithm, along with the EM solver, creates a computational engine for optimization.

The developed computation engine passes the coordinates of each particle during every step in the iteration to the EM solver. Using this dimensional information and solver specific commands, which involve assigning of material properties to each component in the structure, the solver is able to design the structure, assign boundary conditions, assign EM solver parameters, simulate the structure for a particular range, and output the S-parameters obtained from the simulation. These S-parameters are sent back to the MOPSO kernel, which uses these values in order to calculate application-specific fitness functions.

8.4.2 High performance ultra-thin absorber

As mentioned previously, minimization of thickness along with maximization of absorption is often a design criterion for absorbers. A trade-off between these two objectives can be easily achieved using multi-objective particle swarm optimization. Therefore, the following two objective functions f_1 and f_2 are chosen in this work:

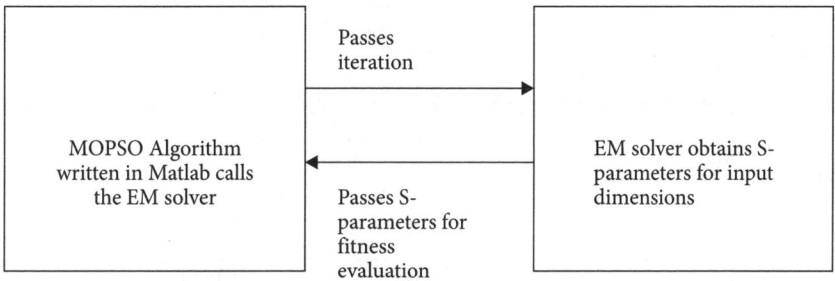

Fig 8.9 Schematic representation of MOPSO based computational engine developed in this project

$$f_1 = S_{11}\big|_{2THz} \tag{8.2}$$

$$f_2 = a + t_d \tag{8.3}$$

where, S_{11} is the scattering parameter, a is the dimension of the substrate of the SRR in the direction of wave propagation and t_d is the dimension of the additional substrate (spacer) in the direction of wave propagation.

The developed MOPSO based computational engine simultaneously optimizes Eqs. 8.2 and 8.3. The parameters assigned in the developed algorithm are given in Table 8.2. Each particle is considered to have six dimensions, viz. outer radius of the external ring (r_{ext}), width of the rings (w), gap length (g), distance between the two rings (d), height of the substrate (h), and dimension of the spacer in the direction of wave propagation (t_d). The dimension of the substrate in the direction of wave propagation, a is taken as

$$a = r_{ext} + 0.2 \tag{8.4}$$

For the simulation, a total of 6 particles are chosen and the algorithm is run for 20 iterations. The resultant *pareto front* obtained is given in Fig. 8.10. These values correspond to coordinates wherein one fitness value can be improved only by degrading the other. The solution at the end of the algorithms run is a collection of all the points on the *pareto front*, and is given in Table 8.3. The total computational time taken for this simulation is 7.8 hours in a *dual core* 4 GB RAM Intel PC.

Table 8.2 MOPSO parameters used for design optimization of ultra-thin THz absorber

Parameters	Value	Use
Wt	1	Inertial constant
$C1$	0.5	Constant1, determine relative pull of *pbest*
$C2$	0.5	Constant 2, determine relative pull of *gbest*
Np	6	Number of particles
Nd	6	Number of dimensions
Nt	20	Number of time steps

Table 8.3 *Pareto front* solutions for the 2 THz absorber (dimensions in μm)

S_{11}	Total thickness	r_{ext}	w	g	d	h	t_d
0.0998	15.7231	5.3	0.4	0.3413	0.2796	0.8049	4.7231
0.3102	12.1697	5.1897	0.2611	0.36	0.2835	0.7966	1.3903
0.3932	11.4972	5.1877	0.3011	0.3464	0.294	0.7993	0.7219
0.3824	11.5145	5.2072	0.3245	0.344	0.2707	0.8028	0.7
0.2536	13.619	5.164	0.2667	0.3977	0.2569	0.7997	2.8911
0.2355	14.0229	5.2016	0.2465	0.3888	0.2699	0.7959	3.2196
0.1774	14.1722	5.1488	0.2267	0.4	0.2639	0.797	3.4747
0.2273	14.1062	5.1845	0.2093	0.395	0.2704	0.7954	3.3372
0.3587	12.0919	5.1871	0.2092	0.3882	0.2631	0.7929	1.3176
0.1477	14.5762	5.1573	0.2	0.4	0.2691	0.7947	3.8616
0.2811	12.9808	5.1778	0.2	0.3851	0.2693	0.7938	2.2252
0.2384	13.7102	5.1754	0.2	0.3806	0.2716	0.7951	2.9594
0.1322	15.2308	5.2246	0.28	0.3761	0.2781	0.7982	4.3816
0.0314	15.9955	5.2556	0.3164	0.3637	0.2849	0.801	5.0843
0.0481	15.9478	5.3	0.3831	0.3318	0.2839	0.8072	4.9478
0.0147	16.2765	5.2974	0.3814	0.3359	0.2862	0.8055	5.2817
0.0165	16.1342	5.3	0.3886	0.3312	0.2832	0.8057	5.1342
0.0547	15.9463	5.3	0.3953	0.3346	0.2847	0.8068	4.9463
0.0665	15.8944	5.3	0.4	0.3343	0.2832	0.8058	4.8944
0.0583	15.9224	5.3	0.4	0.3337	0.28	0.8059	4.9224
0.0738	15.835	5.3	0.4	0.3384	0.2796	0.8058	4.835
0.1051	15.7005	5.3	0.4	0.3359	0.2813	0.8058	4.7005
0.0731	15.8153	5.3	0.4	0.3372	0.2789	0.8053	4.8153
0.0797	15.8341	5.3	0.4	0.3365	0.2807	0.8056	4.8341
0.0862	15.7394	5.3	0.4	0.3411	0.2789	0.8054	4.7394

The designer can chose any value amongst the entries given in Table 8.3 after assigning priority to the fitness functions in her/his design. Therefore, MOPSO is a useful tool in the design of high performance, compact absorbers. Once again, more solutions may be obtained by increasing the number of particles and total number of iterations. However, this will result in an increase in the overall computational time.

8.5   Soft Computing for Active Terahertz Absorber

It is seen that the absorption bandwidth of the SRR based absorber is very small in as evident from the results obtained in the previous section (Figs. 8.7 and 8.8). However, for most applications, high absorption is required over a wide bandwidth. Consequently, the SRR-based absorber designs must be fitted with mechanisms to tune the resonant frequency. From the study in Section 8.2, it is concluded that tuning mechanisms can be classified into four broad categories:

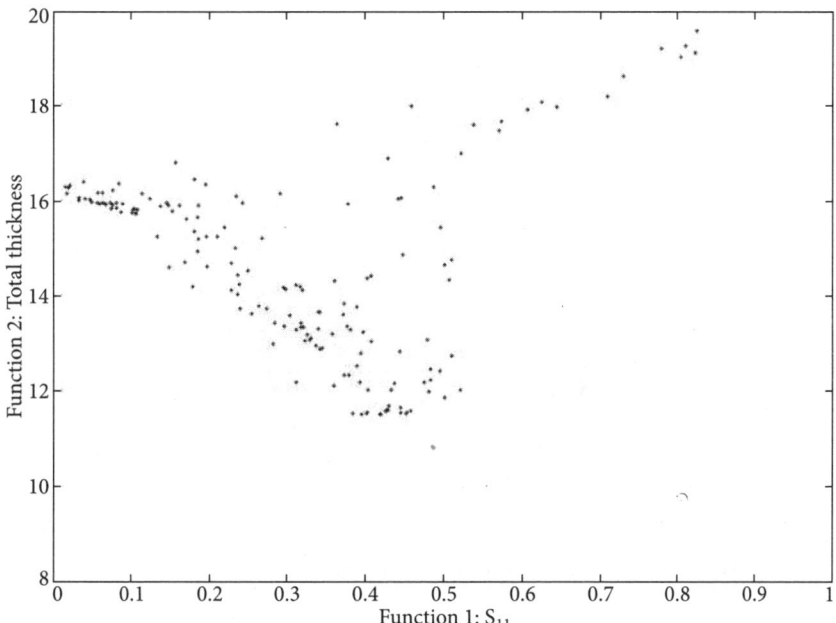

Fig 8.10   Pareto front obtained for 2 THz absorber design

MEMS-based actuation, electrical actuation, photo-excitation, and thermal actuation.

8.5.1 Selection of tuning mechanism

Both thermal excitation and photo-excitation are not feasible from view of integration with detectors as these energies might appear as noise to the detector. Electrical actuation involves changing the properties of the substrate through bias voltages. This requires specialized materials such as barium strontium titanate (BST) to be used as substrate [Bian *et al.*, 2014], etc. The properties of such materials are inconsistent with the contemporary design. Therefore, the MEMS-based tuning method is best suited for this work. From literature, it is seen that the resonant frequency of a circular split ring resonator depends on the relative position of the gaps in the two rings [Saha and Siddiqui, 2009]. When the inner ring is rotated by an angle θ [Fig. 8.11], the resonant frequency f_o can be calculated using Eq. 8.5 through Eqs. 8.6–8.8.

$$f_o = \frac{1}{2\pi\sqrt{2r_o L \dfrac{(\pi+k)^2 - \theta^2}{2(\pi+k)} C_{pul}}} \tag{8.5}$$

where, r_o is the outer radius of the inner ring, L is the total inductance of the SRR (Eq. 4.4), C_{pul} is the per unit length capacitance between the rings, whereas k is calculated as:

$$k = \frac{C_g}{r_o C_{pul}} \tag{8.6}$$

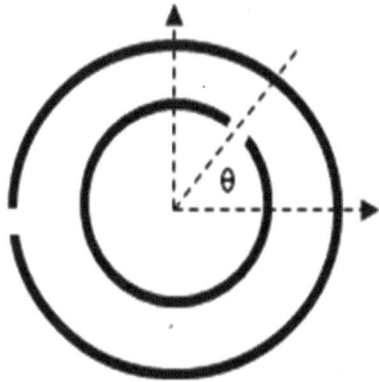

Fig 8.11   Rotation of inner ring by angle θ

where, C_g is the gap capacitance and is given by

$$C_g = \frac{\varepsilon_o wh}{g} \tag{8.7}$$

where, w is the width of the rings, h is the thickness of the substrate and g is the gap length. The values for L and C_{pul} may be calculated from

$$L = 0.005081(2.303 \log_{10} \frac{4l}{d} - \gamma) \tag{8.8}$$

Here, $\gamma = 2.451$ and l and d are wire length and width, respectively.

8.5.2 Implementation of tuning mechanism

From Eq. 8.5, it is seen that as θ (Fig. 8.11) increases, the resonant frequency increases. However, practically, rotating the inner ring by itself is not an easy task. Therefore, this rotational behaviour is obtained by placing switches at discreet angles along the inner ring. When the switch is on, it shortens the rings. When switched off, it behaves as a gap. Only one switch can be switched off at a particular instant. This concept is implemented in the absorber for 1 THz. The results for the same are given in Table 8.4 and Fig. 8.12. Significant absorption was seen after rotation of 90°. Excellent absorption characteristics can be therefore, obtained by this mechanism, over a tunable range of 0.5 THz. For increasing the bandwidth further, an active hybrid absorber array is proposed with unit cell designed for a resonant frequency 0.5 THz apart.

Significant change in the resonant frequency is observed only after 90° rotation of the inner ring. No resonance is observed when the gap of the inner ring is aligned with that of the outer ring, i.e., the no resonance is seen when the angle of rotation is 180°.

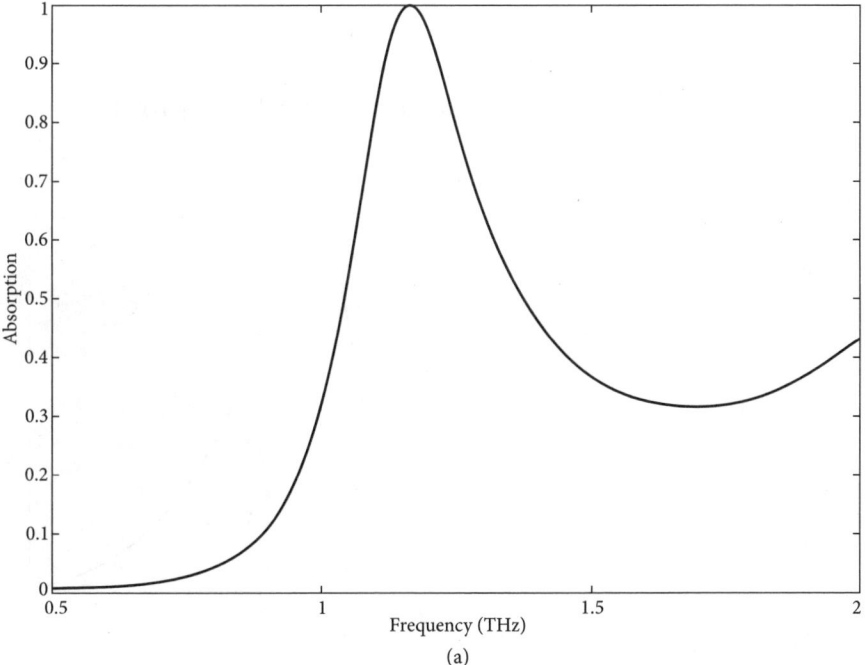

(a)

Fig 8.12(a) Absorption characteristics for different rotation angles: 90°

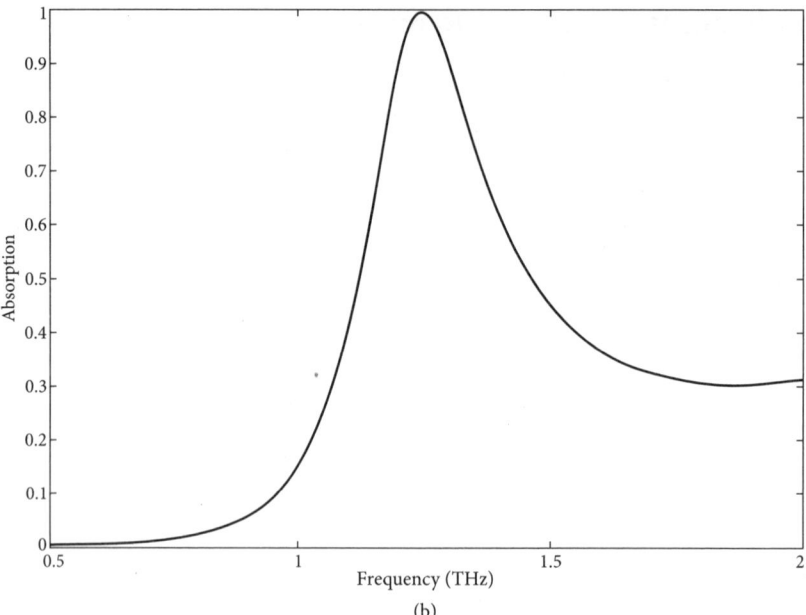

(b)

Fig 8.12(b)   Absorption characteristics for different rotation angles: 120°

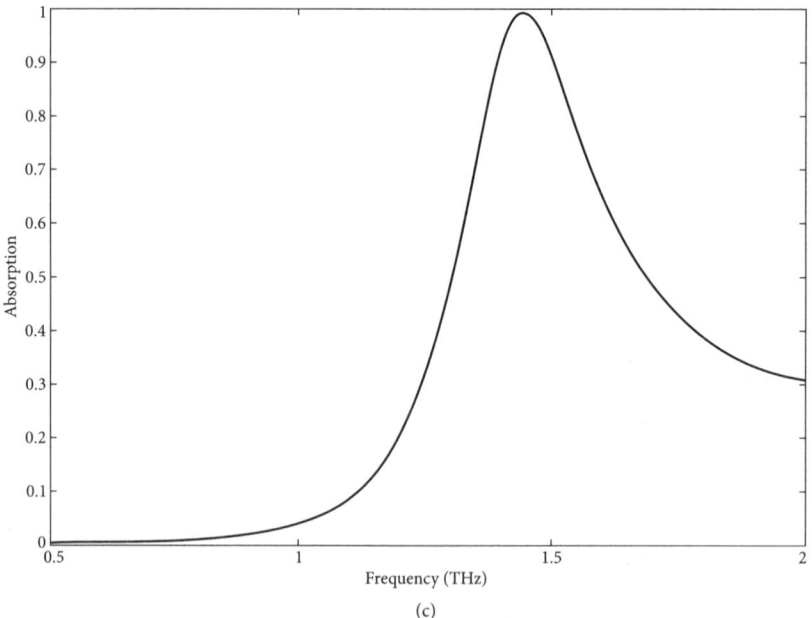

(c)

Fig 8.12(c)   Absorption characteristics for different rotation angles: 135°

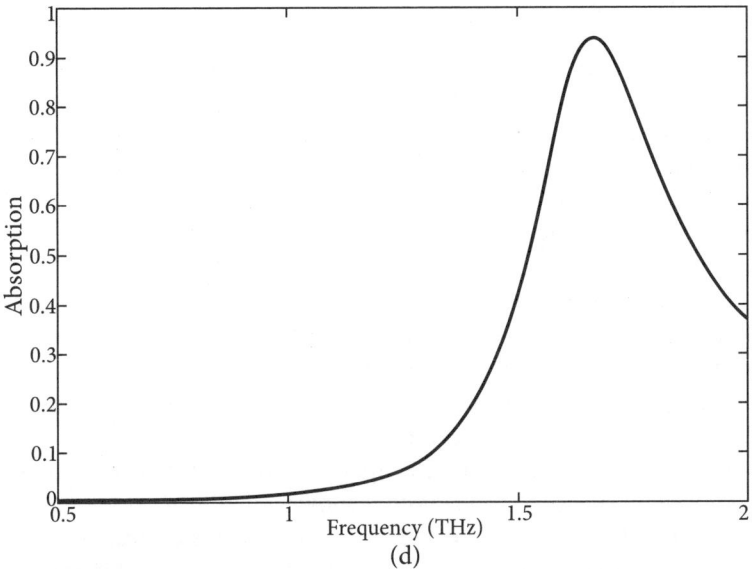

Fig 8.12(d)   Absorption characteristics for different rotation angles: 150°

Table 8.4   Implementation of tuning mechanism for SRR

Angle of Rotation	Resonant Frequency (THz)	Absorption (%)
0°	1	98.73
90°	1.165	100
120°	1.242	99.4
135°	1.444	99.03
150°	1.664	93.73

8.5.3 PSO for design of active absorber array

The above mentioned tuning mechanism is then implemented into a three element absorber array. Here, PSO is used to determine the rotation of each ring in the array for resonance at a particular resonant frequency.

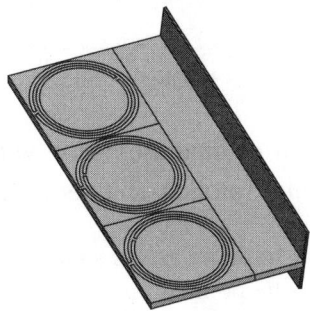

Fig 8.13   Schematic representation of three element absorber array

8.5.3.1 Design procedure

The optimized split ring resonator obtained for 2 THz is used as a unit cell for the three element array as shown in Fig. 8.13. Each unit cell is fitted with switches placed 20° apart. The activation of these switches in a unit cell is taken to be independent of the other unit cells. A PSO-based computational engine is then used to determine the switches that must be activated in each unit cell for resonance at 2 THz. Since there are three unit cells, the dimension of each particle is taken to be 3. Furthermore, the coordinates are allowed to take only integer values from [0, 9] in order to denote the position of the switch in between 0° and 180°. Before passing these values to the EM solver, they are multiplied by 20 (as the switches are place at an angular distance of 20° on the inner ring). The EM solver then creates the structure based on structural parameters obtained in the previous section, rotates each ring by the angle that is given as the input, simulates and obtains the S-parameters, and passes these S-parameter values back into the PSO kernel. The fitness is then determined by using Eq. 8.2.

The computational engine is run for 8 PSO particles and 20 iterations. The variation in the minimum fitness function is stored and plotted as shown in Fig. 8.14. At the end of the algorithm, the best fitness values obtained is found to be S_{11}=0.1476. This corresponds to absorption of 97.82 %. The rotation angles for the rings are found to be 0°, 60°, and 40°, respectively.

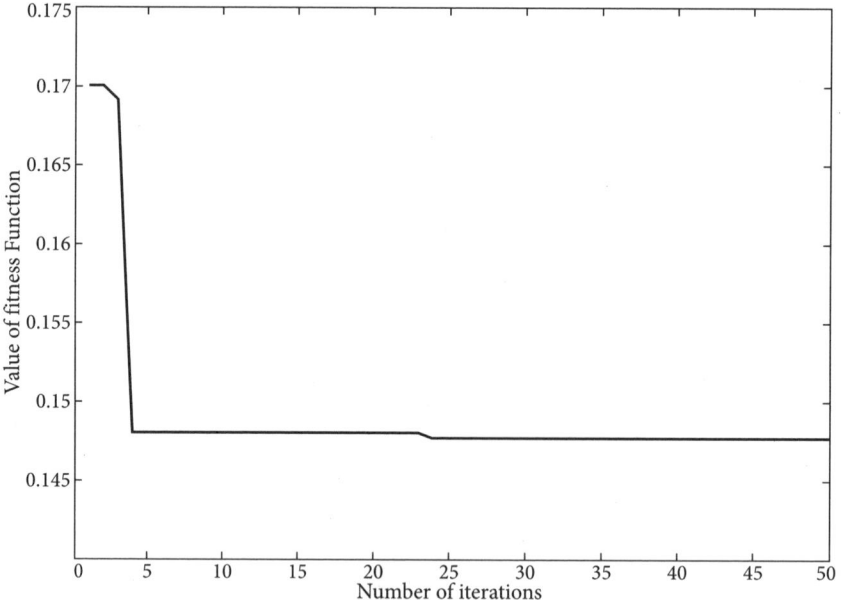

Fig 8.14 Variation of fitness function with iterations for determination of angular rotation of inner rings in 3 element absorber array

The total simulation time taken to arrive at this solution is 12 hours *dual core* 4 GB RAM Intel PC. At this juncture, it must be pointed out that better results may be obtained if the algorithm is allowed to take any angle instead of angles at multiples of 20°. However, from a practical point of view, switches can be placed only at discreet angular intervals. Furthermore, the absorption may be improved by increasing the number of particles and total number of iterations in the PSO code. In this case, the overall computational time will also increase.

8.5.3.2 Concept of adaptive tuning

A database consisting of the rotation angles required for obtaining resonance at a particular frequency can hence be prepared. This database may be then stored in the memory of the processer in the active absorber array system. It can serve as the basis for developing adaptive tunable absorber arrays and can be programmed into a microcontroller. The presence of the database in the memory implies that the angles of rotation for a particular frequency may be obtained quickly without the help of the particle swarm optimization algorithm. This concept is diagrammatically represented in Fig. 8.15.

8.6   Fabrication Sensitivity Analysis

The PSO-based computational engine provides the user with optimized design parameters for high performance absorbers. These solutions are accurate up to four decimal places. However, the fact whether such accuracy may be obtained during actual fabrication, is a factor that must be discussed. In this section, the practicality of the design is verified in terms of fabrication issues.

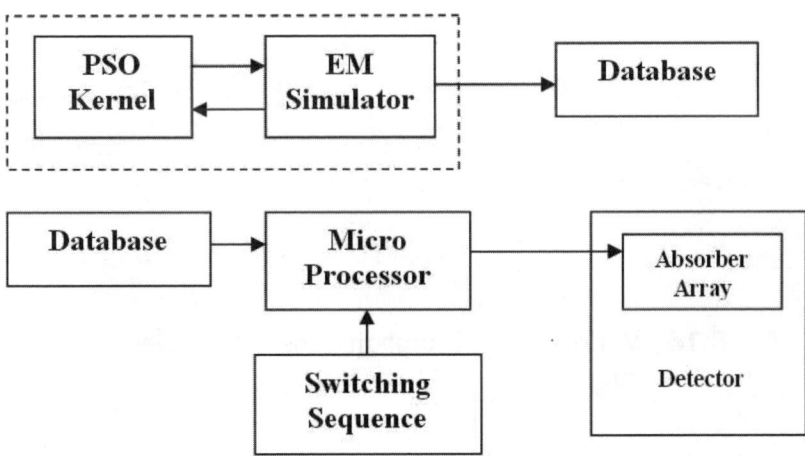

Fig 8.15 Implementation of adaptive tuning

The most commonly used fabrication technique is photolithography. The photolithography equipment currently available can fabricate devices with minimum feature size of 2–2.5 μm. As a result, this technique cannot be used for the fabrication of the optimized absorber as the desired accuracy cannot be achieved. However, these dimensions can be easily realized using electron-beam lithography, and this technique is recommended for fabrication. However, a margin of 5–10% must be allowed for practical deviations from the design.

It is recommended to allow for a 5% shift in the material properties (in this case, permittivity of the substrate), in order to compensate for any impurities in the material. Further, for rugged applications, it is recommended to use aluminium instead of gold as the material for the split ring resonators, since gold is susceptible to wear and tear.

After consultation with material and fabrication experts, it was concluded that the optimized design parameters are within the fabrication limits of electron beam lithography (EBL). Hence, EBL will yield accurate structures. Further, slight errors in the design and/or variations in the material properties should not affect the overall performance of the structure as the absorption will remain greater than 90% within the bandwidth of operation (Fig. 8.16). Therefore, the results obtained from the PSO-based computational engine are practical and can be implemented in order to design high performance, metamaterial absorbers.

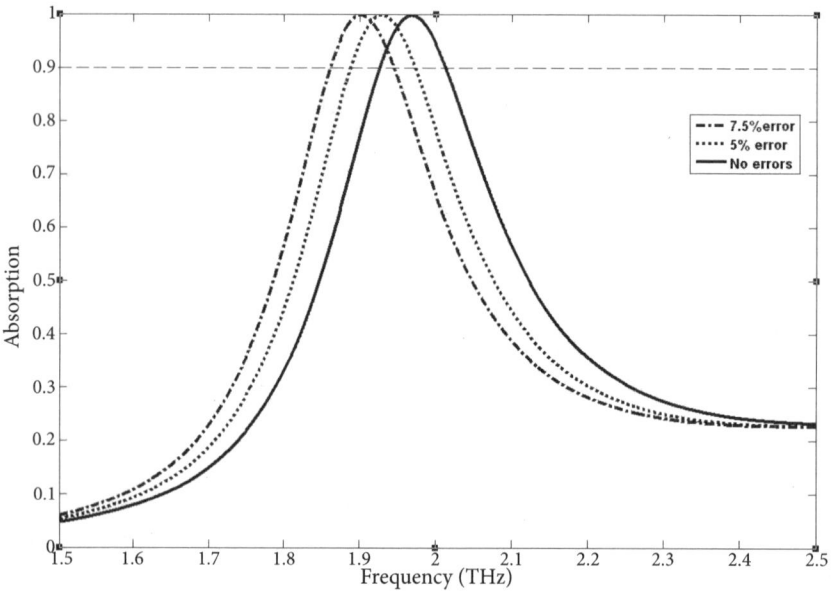

Fig 8.16 Variation of absorption considering tolerance during actual fabrication

8.7 Summary

This chapter provides an in-depth knowledge of capabilities of soft computing in complicated designs through development of optimization-based computational engines that can integrate the optimization tool along with commercially available technical tools. Design of active terahertz metamaterial absorber with adaptive tuning mechanism is considered as an example here. Towards this, a MOPSO based computational engine, which integrates a MOPSO kernel with a commercial EM solver, is developed so as to optimize the structural parameters of the absorber for high performance.

This computational engine, MOPSO eliminates the need for manually determining the best structural parameters and is especially useful when the design equations are not well identified. In addition to structural optimization, the computational engine is also used to determine the optimum inner ring rotational angles for a three-element circular SRR absorber array. This engine is capable of determining the rotational angles for different resonant frequencies. These values constitute a database, which may be used for the rapid, adaptive tuning of the absorber array.

References

Alves, F., B. Kearney, D. Grbovic, and G. Karunasiri, "Narrowband terahertz emitters using metamaterial films," *Optics Express,* vol. 20, no. 19, pp. 21025–21032, Sep. 2012.

Basiry, R., H. Abiri, and A. Yahaghi, "Wide-band, stop-band and multi-band metamaterial filter design in terahertz frequencies," *Proceedings of Second Conference on Millimeter-Wave and Terahertz Technologies (MMWaTT),* pp. 33–36, Dec. 2012.

Bian, Y., C. Wu, H. Li, and J. Zhai, "A tunable metamaterial dependent on electric field at terahertz with barium strontium titanate thin film," *Applied Physics Letters,* vol. 104, pp. 042906-1–042906-4, Jan. 2014.

Chen, W. C., J. J. Mock, D. R. Smith, T. Akalin, and W. J. Padilla, "Controlling gigahertz and terahertz surface electromagnetic waves with metamaterial resonators," *Physical Review X,* vol. 1, pp. 021016(1)–021016(6), 2011.

Chowdhury, D. R., R. Singh, J. F. O'Hara, H. T. Chen, A. J. Taylor, and A. K. Azad, "Photo-doped silicon in split ring resonator gap towards dynamically reconfigurable terahertz metamaterial," *CLEO Technical Digest,* 2p., 2012.

Ekmekci, E., K. Topalli, T. Akin, and G. T.-Sayan, "A tunable multi-band metamaterial design using micro-split SRR structures," *Optics Express,* vol. 17, no. 18, pp. 16046–16058, Aug. 2009.

Ferguson, B., and X. C. Zhang, "Materials for terahertz science and technology," *Nature Materials,* vol. 1, pp. 26–33, Sept. 2009.

Headland, D., W. Withayachumnankul, M. Webb, and D. Abbott, "Beam deflection lens at terahertz frequencies using a hole lattice metamaterial," *38th International Conference on Infrared, Millimeter, and Terahertz Waves (IRMMW-THz),* pp. 1–2, Sep. 2013.

Horestani, A. K., W. Withayachumnankul, A. Chahadih, A. Ghaddar, M. Zhar, D. Abbott, and T. Akalin "Metamaterial-inspired bandpass filters for terahertz surface waves on Goubau Lines," *IEEE Transactions on Terahertz Science and Technology,* vol. 3, no. 6, pp. 851–858, Nov. 2013.

Hwu, S., K. B. deSilva, and C. T. Jih, "Terahertz (THz) wireless systems for space applications," *Proceedings of IEEE Sensors Applications Symposium,* pp. 171–175, Feb. 2013.

Kan, K., J. Yang, A. Ogata, T. Kondoh, K. Norizawa, Y. Yoshida, M. Hangyo, R. Kuroda, and H. Toyokawa, "Terahertz-wave generation using metamaterial and femtosecond electron bunch," *37th International Conference on Infrared, Millimeter, and Terahertz Waves (IRMMW-THz),* pp. 1–2, Sep. 2012.

Kearney, B. T., *Enhancing microbolometer performance at terahertz frequencies with metamaterial absorbers*, Doctorate of Philosophy dissertation, 69 p., Naval Postgraduate School, 2013.

Kowerdziej, R., M. Olifierczuka, B. Salskib, and J. Parkaa, "Tunable negative index metamaterial employing in-plane switching mode at terahertz frequencies," *Liquid Crystals*, vol. 39, no. 7, pp. 827–831, Jul. 2012.

Kozlov, D. S., M. A. Odit, I. B. Vendik, Y.-G. Roh, S. Cheon, and C. -W. Lee, "Tunable terahertz metamaterial based on resonant dielectric inclusions with disturbed Mie resonance," *Applied Physics A: Material Science and Processing*, pp. 465–470, Dec. 2011.

Lu, M., W. Li, and E. R. Brown, "High-Order THz bandpass filters achieved by multilayer complementary metamaterial structures," *35th International Conference on Infrared Millimeter and Terahertz Waves (IRMMW-THz)*, pp. 1–2, Sep. 2010.

Maagt, P., "Terahertz technology for space and earth applications," *International Workshop on Antenna Technology: Small and Smart Antenna, Metamaterials and Applications*, pp. 111–115, 2007.

Nemec, H., P. Kuzel, F. Kadlec, C. Kadlec, R. Yahiaoui, and P. Mounaix, "Tunable terahertz metamaterials with negative permeability," *Physical Review Letters*, vol. 79, pp. 241108-1–241108-4, May. 2009.

Neu, J., B. Krolla, O. Paul, B. Reinhard, R. Beigang, and M. Rahm, "Metamaterial based gradient index lens with strong focusing in the THz frequency range," *Optics Express*, vol. 18, issue 26, pp. 27748–27757, 2010.

Ozbey, B., and O. Aktas, "Continuously tunable terahertz metamaterial employing magnetically actuated cantilevers," *Optics Express*, vol. 19, no. 7, pp. 5741–5752, Mar. 2011.

Padilla, W. J., A. J. Taylor, C. Highstrete, M. Lee, and R. D. Averitt, "Dynamical electric metamaterial response at terahertz frequencies," *Proceedings of the 15th International Conference, Pacific Grove, USA*, pp. 642–644, Aug. 2006.

Padilla, W. J., "Metamaterial devices for the terahertz gap," *Asia Pacific Microwave Conference, APMC* 2009, pp. 1297–1298, Dec. 2009.

Paul, O., C. Imhof, B. Lägel, S. Wolff, J. Heinrich, S. Höfling, A. Forchel, R. Zengerle, R. Beigang, and M. Rahm, "Electrically tunable metamaterial for polarization-independent terahertz modulation," *Conference on Lasers and Electro-optics*, pp. 1–2, Jun. 2009.

Pratibha, R., K. Park, I. I. Smalyukh, and W. Park, "Tunable optical metamaterial based on liquid crystal-gold nanosphere composite," *Optics Express*, vol. 17, no. 22, pp. 19459–19469, Oct. 2009.

Rout, S., D. Shrekenhamer, S. Sonkusale, and W. Padilla, "Embedded HEMT/Metamaterial composite devices for active terahertz modulation," *23rd Annual Meeting of the IEEE Photonics Society*, pp. 437–438, Nov. 2010.

Sabah, C. and H. G. Roskos, "Periodic array of chiral metamaterial-dielectric slabs for the application as terahertz polarization rotator," *XXXth URSI General Assembly and Scientific Symposium*, pp 1–4, Aug. 2011a.

Saha, C. and J. Y. Siddiqui, "Estimation of the resonance frequency of the conventional and rotational circular split ring resonators," *Proceedings of Applied Electromagnetics Conference (AEMC)*, pp. 1–3, Dec. 2009.

Shadrivov, I. V., S. K. Morrison, and Y. S. Kivshar, "Tunable split-ring resonators for nonlinear negative-index metamaterials," *Optics Express*, vol. 14, no. 20, pp. 9344–9349, Sep. 2006.

Sheng, L. J. and Z. X. Li, "Terahertz wave modulator based on metamaterial," *Journal of Physics: Conference Series*, vol. 276, pp. 012213 (1)–012213 (3), 2011*a*.

Siegel, P. H., "Terahertz technology in biology and medicine," *IEEE Transactions on Microwave Theory and Techniques*, vol. 52, no. 2, pp. 2438–2447, Oct. 2004.

Smith, D. R., J. B. Pendry, and M. C. K. Wiltshire, "Metamaterials and negative refractive index," *Science*, vol. 305, pp. 788–792, Aug. 2004.

Strikwerda, A. C., H. Tao, E. A. Kadlec, K. Fan, W. J. Padilla, R. D. Averitt, E. A. Shaner, and X. Zhang, "Metamaterial based terahertz detector," *36th International Conference on Infrared, Millimeter and Terahertz Waves (IRMMW-THz)*, pp. 1–2, Oct. 2011.

Tao, H., A. Strikwerda, C. Bingham, W. J. Padilla, X. Zhang, and R. D. Averitt, "Dynamical control of terahertz metamaterial resonance response using bimaterial cantilevers," *PIERS Proceedings*, pp. 870–873, Jul. 2008.

Vendik, I. B., O. G. Vendik, M. A. Odit, D. V. Kholodnyak, S. P. Zubko, M. F. Sitnikova, P. A. Turalchuk, K. N. Zemlyakov, I. V. Munina, D. S. Kozlov, V. M. Turgaliev, A. B. Ustinov, Y. Park, J. Kihm, and C. W. Lee, "Tunable metamaterial structures for controlling THz radiation," *IEEE Transactions on Terahertz Science and Technology*, vol. 2, no. 5, pp. 538–549, Sep. 2012.

Yang, T., X. Li, and W. Zhu, "A tunable metamaterial absorber employing MEMS actuators in THz regime," *8th IEEE International Conference on Nano/Micro Engineered and Molecular Systems (NEMS)*, pp. 829–832, 2013.

Zhang, W., W. M. Zhu, H. Cai, P. Kropelnicki, A. B. Randles, M. Tang, H. Tanoto, Q. Y. Wu, J. H. Teng, X. H. Zhang, D. L. Kwong, and A. Q. Liu, "A tunable MEMS THz waveplate based on isotropicity dependent metamaterial," *Transducers*, pp. 538–541, June 2013.

Zhu, W. M., A. Q. Liu, T. Bourouina, D. P. Tsai, J. H. Teng, X. H. Zhang, G. Q. Lo, D. L. Kwong, and N. I. Zheludev, "Micro-electromechanical Maltese-cross metamaterial with tunable terahertz anisotropy," *Nature Communications*, 6p., Dec. 2012.

CHAPTER

9

Soft Computing Based CAD Packages for EM Applications

This chapter covers the use of soft computing for creating various CAD packages. These packages deliver a range of features including efficiency, quick response as well as an easy-to-use user interface that allows even users not well versed with soft computing to arrive at optimized solutions for their applications. In the chapter, the development of two such CAD packages is discussed: one for the design of metamaterial split ring resonators and the other for path loss prediction in rural and urban environments. The CAD package for metamaterial structures uses PSO for design optimization whereas the path loss prediction CAD package uses neural network. Soft computing techniques based CAD package are gaining momentum in the field of electromagnetic (EM) applications because of their properties such as global optimization, quick response and accuracy [Mishra and Patnaik, 1993].

9.1   CAD Package for Metamaterial Structures

Metamaterials are artificially designed EM structures that acts as an effective medium with negative refractive index at a desired frequency of operation. Split ring resonators (SRR) are proven metamaterial structures for various applications such as metamaterial based antennas, radar absorbing materials, THz absorbers for biomedical applications, frequency selective surfaces, invisibility cloaks [Kwon *et al.*, 2007; Jin and Samii, 2007; Goudos and Sahalos, 2006; Ivsic *et al.*, 2010], etc. Due to the inherent narrowband operation of the split-ring resonators, it is critical to ensure that the resonant frequency of the same is equal to that of the application.

Therefore, a CAD package is developed here to obtain the optimized structural parameters for different configurations of the metamaterial SRR for a desired frequency of operation. This CAD package uses *particle swarm optimization* (PSO), which is based on the movement and intelligence of swarms as discussed in Chapter 2, for optimizing difficult multidimensional discontinuous problems. The *equivalent circuit analysis* (ECA) method is used here as an electromagnetic solution tool for SRR design optimization. Hence PSO in conjunction with ECA method provides optimal structural parameters of the different configurations of SRRs.

9.1.1 Equivalent circuit analysis of square SRR

The analysis of metamaterial structures is carried out using equivalent circuit analysis method where the metamaterial SRR is represented as equivalent circuit of lumped elements. Several metamaterial configurations, such as single ring, double ring, and triple ring square SRR are considered. In addition to these structures, circular SRR is also analyzed. As discussed in Chapter 4, the designed SRR can be used as a substrate or superstrate for performance enhancement of an antenna. Various configurations such as double ring and triple ring SRR are studied in order to develop a robust CAD package. The physical dimensions of commonly used resonators are required to be of the order of λ/20 in order to achieve the desired permeability at microwave frequencies [Bilotti *et al.*, 2007a]. These limitations can be reduced by magnetic inclusions, which can be achieved through the design of multiple ring split-ring resonators.

The CAD package developed here provides the designer with the structural parameters of the SRR (length or radius, width and gap between the rings) for a desired resonant frequency and substrate permittivity. The equivalent circuit analysis of these SRR configurations is given below.

Square SRR is a metamaterial structure, which consists of square shaped ring with a gap. This structure is generally printed on a dielectric substrate. Here, it is assumed that length of the square strip is *a*, strip width is *w*, the separation between the adjacent rings is *d*, and length of the split is *g*, and that *a*, *w*, *d*, and *g* are same for all the rings in a structure. In this section, the equivalent circuit analysis of single ring SRR as well as multiple ring SRRs is reported. The schematic diagram of various configuration of square SRR are shown in Fig. 9.1.

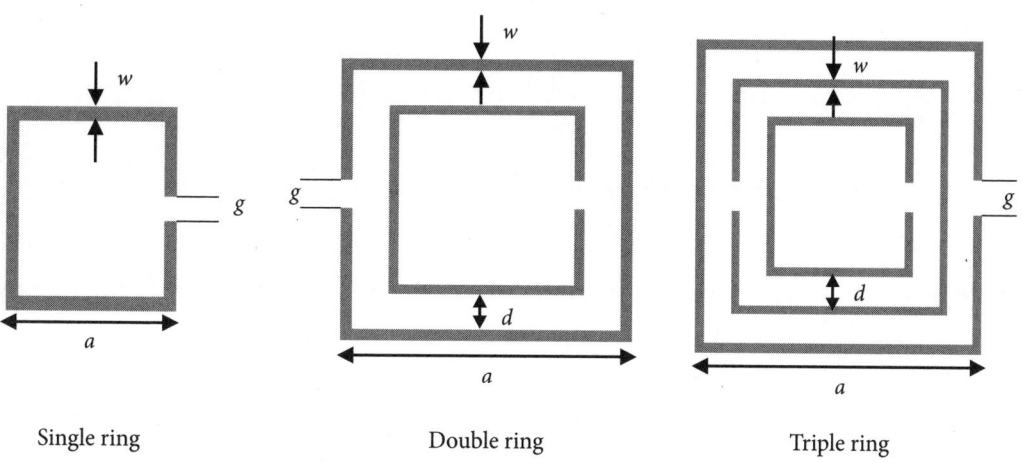

Single ring   Double ring   Triple ring

Fig 9.1   Schematic of various configurations of SRRs

In order to achieve the desired resonant frequency, the expressions for the inductance and distributed capacitances of the circuit model should be derived. The dielectric substrate, upon which the multiple split-ring resonators are printed, does not affect the inductance; however, it affects the capacitance of the equivalent circuit.

The electromagnetic behaviour of the lossless multiple split-ring resonators is described using the equivalent LC series circuit. The value of inductance and capacitance of the multiple split-ring resonators can be determined using the equivalent circuit given in Fig. 9.2 [Bilotti et al., 2007b]. In the proposed equivalent circuit analysis (ECA) model the high-order effects are neglected. Mutual capacitances between non-adjacent rings and mutual inductances between the parallel conducting strips are completely neglected while the split capacitances are assumed to have only minimal impact in the resonant frequency computation.

From the equivalent circuit analysis, resonant frequency of the split ring resonator is given by

$$f_r = \frac{1}{2\pi\sqrt{LC_s}}$$
(9.1)

where, L is effective inductance and C_s is gap capacitance, which are dependent on structural parameters of square SRR. The expressions for L and C_s for the different configurations of square SRR is given in Table 9.1.

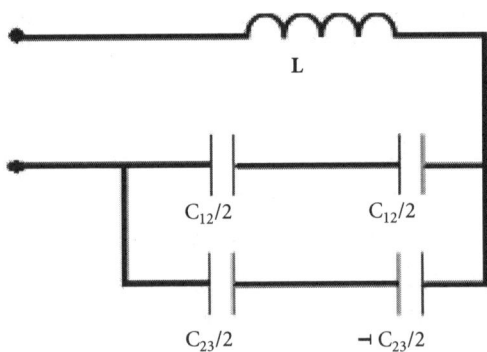

Fig 9.2 Equivalent circuit of a typical square SRR

Table 9.1 The effective inductance and gap capacitance for different configurations of square SRR

Type of square SRR	Effective Inductance, L	Gap Capacitance Cs
Single ring	$L = 2L1 + L2 + 2L3$ (Mukhletely et al. 2010)	$C_s = \dfrac{Kct\varepsilon_0}{g}$

Double ring	$L = \dfrac{4.86\mu_0}{2}(a-w-d)\left[\ln\left(\dfrac{0.98}{\rho}\right)+1.84\rho\right]$	$C_S = \left(a-\dfrac{3}{2}(w+d)\right)C_{pul}$
	Filling factor, $\rho = \dfrac{w+d}{a-w-d}$	
Triple ring	$L = \dfrac{4.86\mu_0}{2}(a-2(w+d))\left[\ln\left(\dfrac{0.98}{\rho}\right)+1.84\rho\right]$	$C_S = (2(a-5)(w+d))C_{pul}$
	Filling factor, $\rho = \dfrac{2(w+d)}{a-2(w+d)}$	

where, K : Correction factor

c : Free space velocity ε_0: Free-space permittivity

t : Thickness of square SRR g : Gap spacing in the square SRR

C_{pul} is the per-unit-length capacitance between the rings

$$C_{pul} = \varepsilon_0\varepsilon_{eff}\frac{K\left(\sqrt{1-k^2}\right)}{K(k)}, \tag{9.1a}$$

ε_{eff} is the effective dielectric constant

$$\varepsilon_{eff} = \frac{\varepsilon_r + 1}{2} \tag{9.1b}$$

$K(k)$ denotes the incomplete elliptical integral of the first kind with k expressed as

$$k = \frac{d}{d+2w} \tag{9.1c}$$

As shown in the equivalent circuit analysis, the resonant frequency of the square SRR depends on the structural parameters. One example for the variation of the resonant frequency w.r.t. to the side length of ring for design of a single ring square SRR is shown in Fig. 9.3. Similarly the resonant frequency vs. the side length of ring for design of a double ring square SRR and a triple ring square SRR is shown in Figs. 9.4 and 9.5.

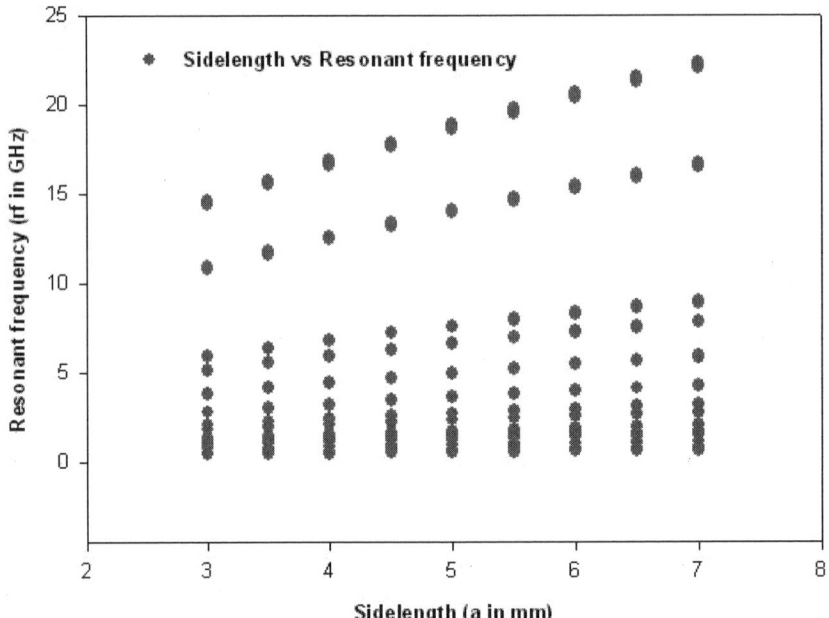

Fig 9.3 The values of side length of single ring square SRR at various frequencies

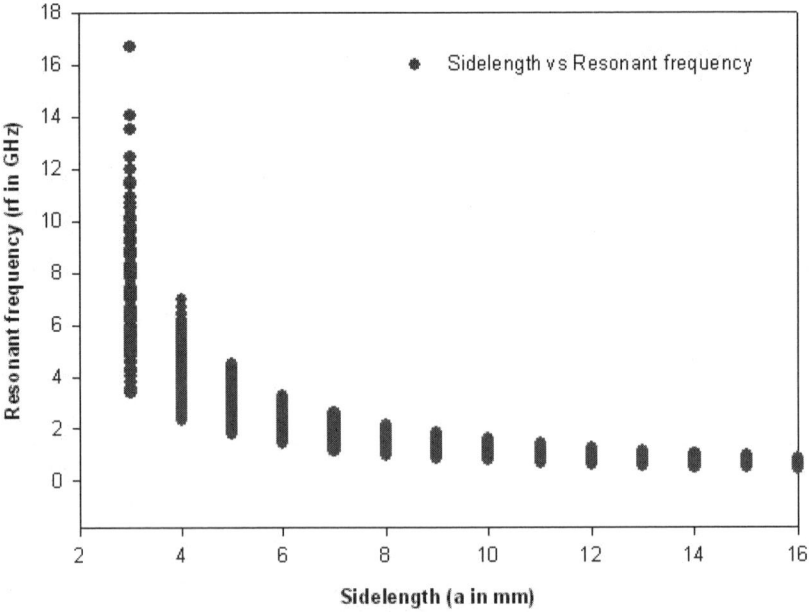

Fig 9.4 The values of side length of double ring square SRR at various frequencies

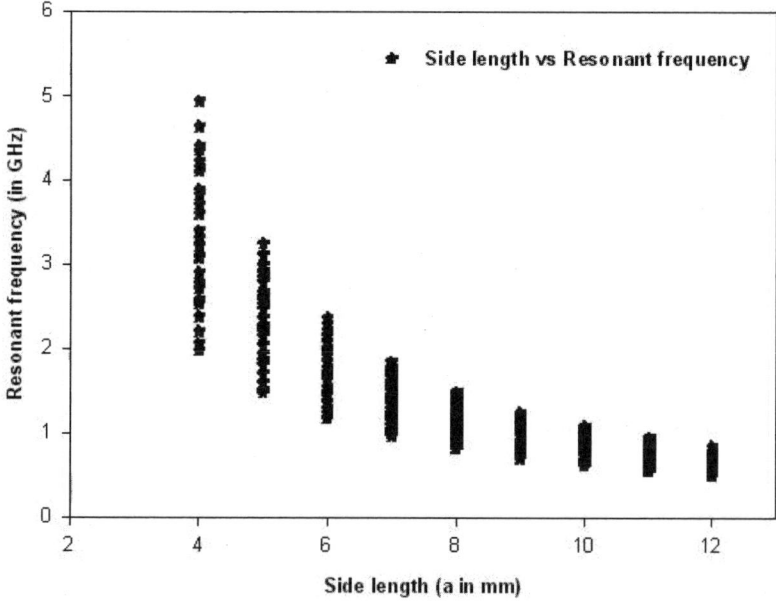

Fig 9.5 The values of side length of triple ring square SRR at various frequencies

9.1.2 Equivalent circuit analysis of circular SRR

The schematic diagram of a circular SRR with the dimensions is shown in Fig. 9.6a where r_{ext} is the external radius, w denotes the width of rings, d is the gap present between the rings and s represents the width of the split in the ring. In this method, the distributed network is converted to lumped network (Fig. 9.6b) to carry out the analysis [Baena *et al.*, 2005].

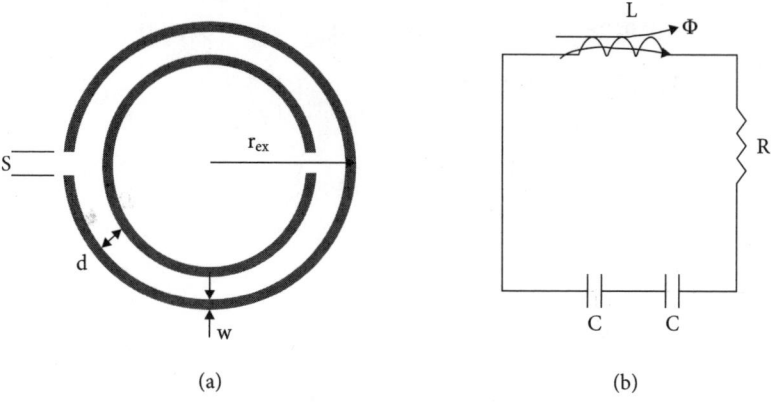

(a)   (b)

Fig 9.6 (a) Schematic of circular SRR (b) Equivalent circuit of circular SRR

The resonant frequency for circular SRR is given by

$$f_0 = \frac{1}{2\pi\sqrt{LC_S}} \tag{9.2}$$

If $r < 5.2$ then the resonant frequency is [Pradeep *et al.*, 2011]

$$f_0 = \frac{1}{2\pi\sqrt{LC_S}} + \frac{5.2 - r_{ext}}{2} \tag{9.3}$$

If $r > 5.2$ then the resonant frequency is

$$f_0 = \frac{1}{2\pi\sqrt{LC_S}} \tag{9.4}$$

where, L and C are the inductance and capacitance of the SRR, respectively. The expressions for L and C are given below.

The inductance L is given by

$$L = 2.57 e^{\frac{-w_3}{\sqrt{2}}} \left(\pi r_{ext} - 2.2 d_1 - \frac{\pi}{2} \right) \tag{9.5}$$

The capacitance C is given as:

$$C = 0.217 + \left\{ \begin{array}{l} \left[0.059 \left(2r_{ext} + \varepsilon_r - 5 \right) \right] \left(0.437 w_1 - 0.317 w_2^2 + 0.07 w_2^3 \right) \\ \left(3.3367 e^{-3.2 d_1} - 0.1955 e^{-0.47 h} \right) \end{array} \right\}$$

$$+ \left(0.05 \varepsilon_r - 0.218 \right) + \left(\frac{0.599 h}{0.0248 + h} - 0.599 \right) \tag{9.6}$$

where, for $d < 1$ mm, $w_1 = w$, $w_2 = w$, $w_3 = w$ and $d_1 = d$.

For $d > 1$ mm, $w_1 = w/3$, $w_2 = 1.414\, w$, $w_3 = w/1.414$ and $d_1 = 1.414\, d$.

Hence, it is seen that the structure of the ring decides its resonant frequency. Figure 9.7 shows the variation of the resonant frequency w.r.t. dimension of the external ring.

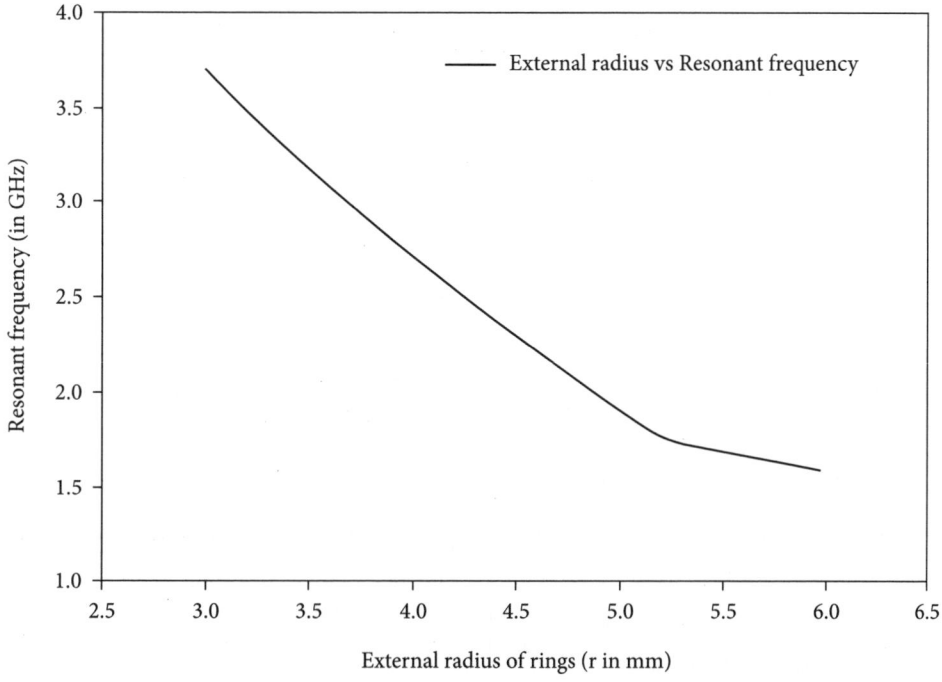

Fig 9.7 The values of external radius of circular SRR obtained at various frequencies

9.1.3 Development of CAD package using PSO

The CAD package designed here is based on lumped element circuit models. The efficiency of any CAD model depends on the computational time and accuracy. The developed CAD package uses PSO, a global optimization technique, in conjunction with the equivalent circuit analysis as EM solver. This combination makes the CAD package more efficient and accurate compared to the other conventional optimizers. The PSO algorithm used here is the same as that described in Chapter 2.

Implementation of the PSO begins with defining the solution space and the fitness function to be optimized. In this optimization problem the fitness function is considered as

$$f_{err} = \frac{|f_d - f_c|}{f_d} \tag{9.7}$$

where, f_d is the desired frequency and f_c is the frequency arrived at by the equivalent circuit analysis.

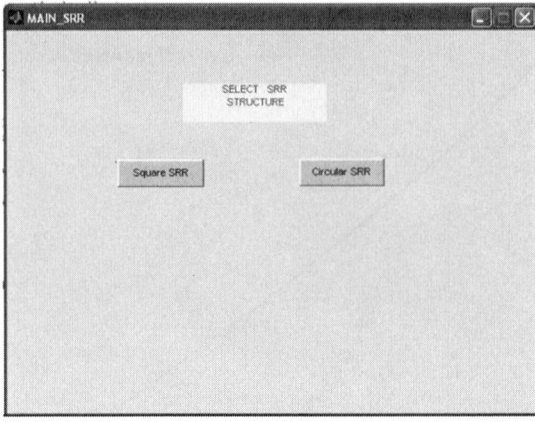

(a)

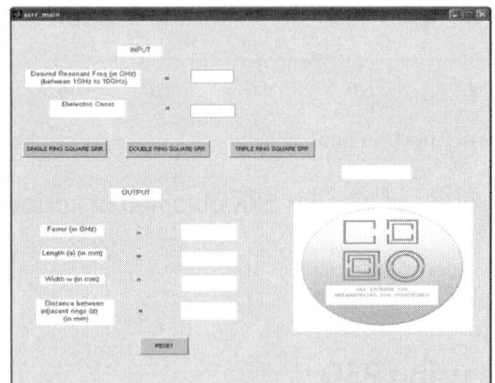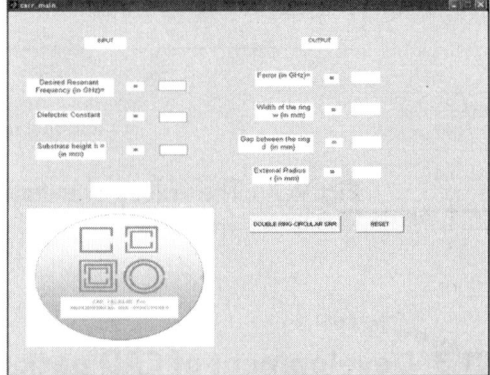

Fig 9.8 Graphical user interface for CAD package (a) Main interface (b) Input interface for square SRR (c) Input interface for circular SRR

To make the CAD package user friendly and efficient, a graphical user interface has been developed (Fig. 9.8), which takes the desired resonant frequency and dielectric constant of the substrate as input and the structural parameters of the SRR as output.

The CAD package is developed using MATLAB, and provides a global solution considering all the structural parameters such as length, width, thickness, as well as dielectric constant of the substrate material. The CAD model optimizes these values in order to obtain an SRR with resonance at the desired frequency.

9.1.4 Optimization of metamaterial structures

The developed GUI takes the resonant frequency and dielectric constant of the substrate as the input and provides the corresponding design parameters as the output. These designs are verified using commercial FEM based software to check the accuracy of the optimization technique.

9.1.4.1 Square SRR

Three configurations of square SRR structures are included in the CAD package namely single ring square SRR, double ring square SRR and triple ring square SRR. The design parameters obtained for all these configurations for a desired resonant frequency and with a known dielectric constant of the substrate are given below:

Table 9.2 PSO extracted design parameters for the three configurations of square SRR

Metamaterial Structure	Dielectric constant of substrate	Desired resonant frequency	PSO extracted design parameters		
			Length a, in mm	Width w, in mm	Distance between adjacent rings d, in mm
Single ring SRR	2.3	4	9	0.3	0.4
Double ring SRR	3.86	7.5	3	0.7	0.4
Triple ring SRR	2.84	6.23	3	0.5	0.25

Screenshots of the developed CAD package for square SRR are shown in Fig. 9.9.

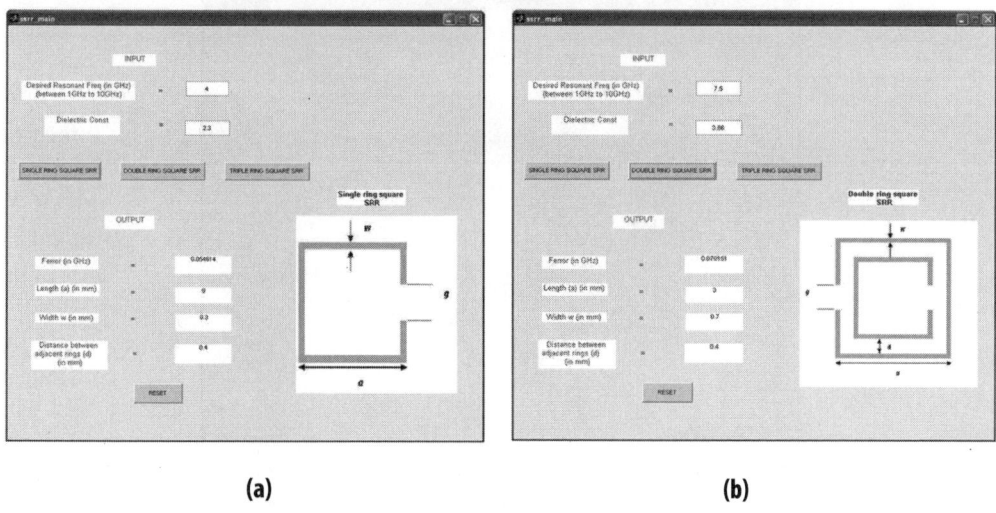

(a)   (b)

Fig 9.9(a & b) Output graphical user interface for (a) Single ring SRR, (b) Double ring SRR

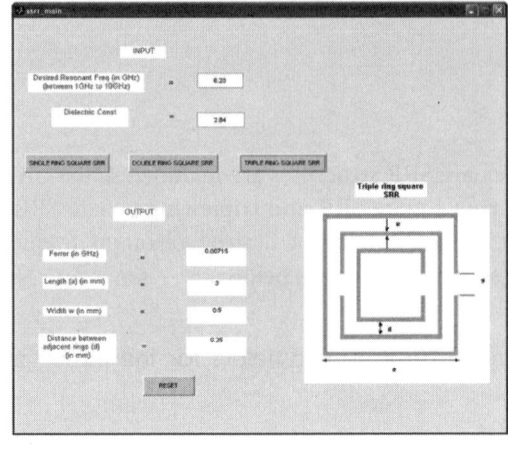

(c)

Fig 9.9(c) Output graphical user interface for: Triple ring SRR

9.1.4.2 Circular SRR

Further, the same CAD package also yields circular SRR configurations (Fig. 9.10). The design parameters are extracted for a desired frequency of 8.25 GHz and substrate dielectric constant of 4.23 with a substrate height of 1.56 (mm) are given as: width of the ring, $w = 0.45$ mm; gap between the ring, $d = 1.1$ mm; external radius, $r = 2$ mm.

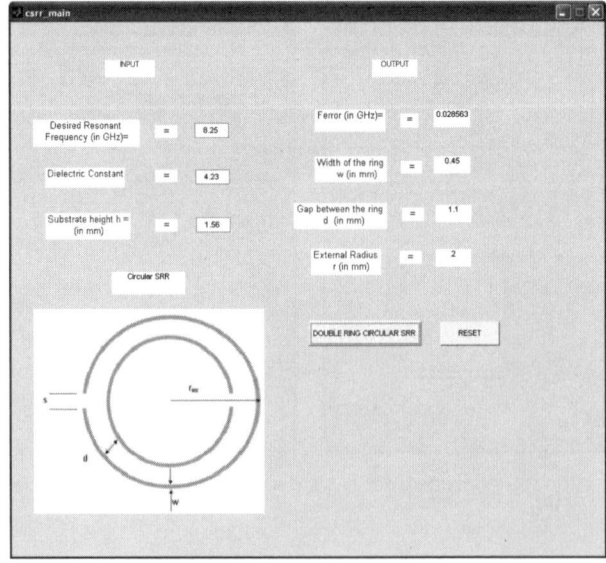

Fig 9.10 Output graphical user interface for circular SRR

9.1.4.3 *Comparison of PSO and GA*

A comparative study of PSO and GA for this same problem has been carried out. It has been observed that the accuracy of GA is less than PSO because in GA the binary coded data is considered [Fan *et al.*, 2007]. The design parameters for a circular SRR at desired resonant frequency 5 GHz and dielectric constant 4, are extracted using both GA and PSO and is given in Table 9.3.

Table 9.3 PSO extracted design parameters for the three configurations of square SRR

Optimization Techniques	Accuracy (f_{err})	Gap between Rings d (in mm)	Width of rings w (in mm)	External radius r (in mm)
GA	0.2970	0.6	0.2	2.2
PSO	0.000011	0.4	0.2	2.35

9.1.5 Applications of the CAD package

The developed CAD package is further validated using FEM-based simulation software and has been used by the authors for various applications such as metamaterial fractal antenna [Choudhury *et al.*, 2013*a*], metamaterial absorber [Choudhury *et al.*, 2013*b*], etc. One validation example of the CAD package is described in this section through simulation studies.

Let us assume that the objective is to design a metamaterial double ring square SRR having resonant frequency of 9.6 GHz. Let the dielectric constant of the substrate be 4.4. The CAD package provides the outputs as side length, a = 2.8 mm, width of the ring, w = 0.3 mm, and gap in the rings, g = 0.3 mm. After obtaining the structural parameters, the simulation studies are carried out (Fig. 9.11) and the corresponding permittivity and permeability are extracted. The simulation results (Figs. 9.12 and 9.13) show that the designed metamaterial square double ring SRR has negative permeability from 9.35 GHz to 9.94 GHz, which fulfils the design objective.

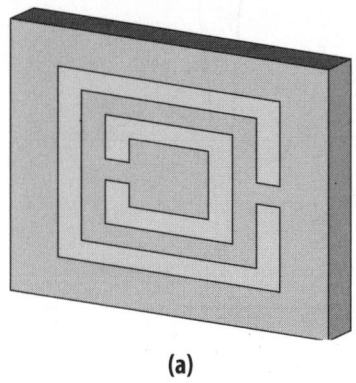

(a)

Fig 9.11(a) PSO optimized double ring square SRR designed using FEM solver,

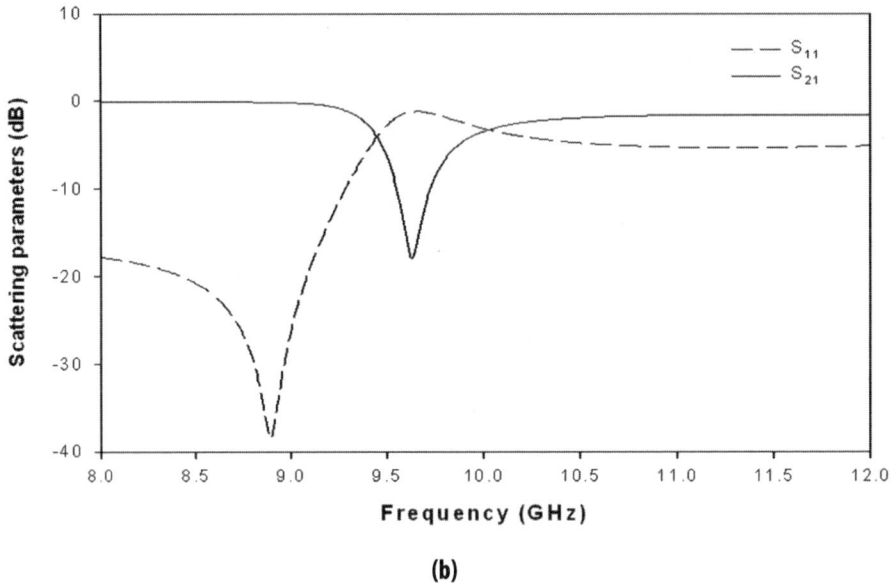

(b)

Fig 9.11(b)   Scattering parameters of the modeled double ring square SRR

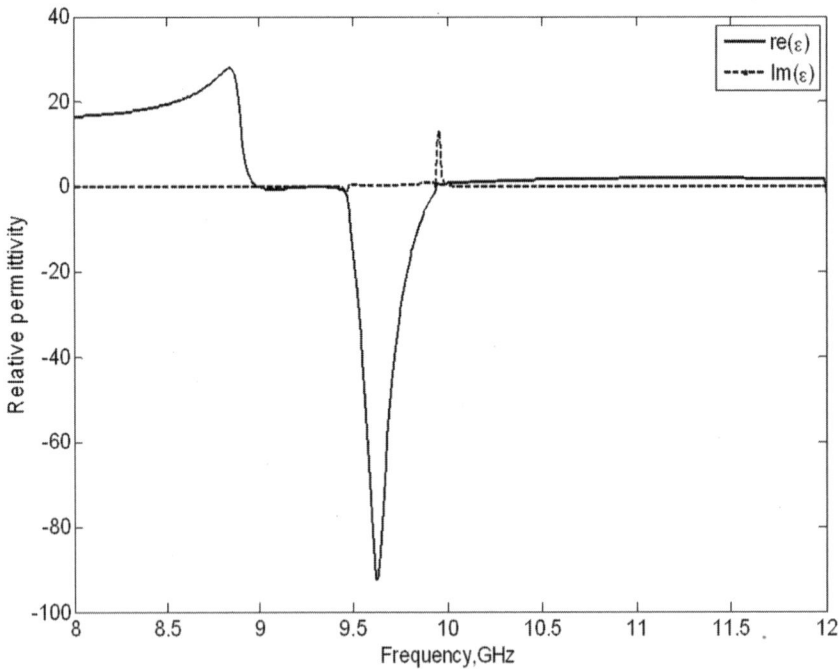

Fig 9.12   Extracted permittivity of the modeled double ring square SRR

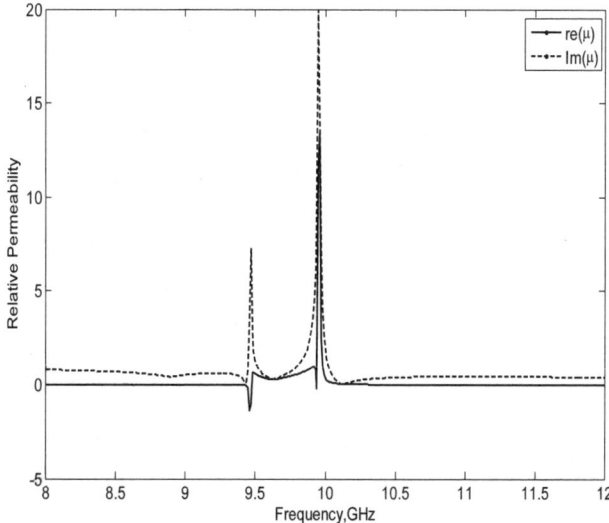

Fig 9.13 Extracted permeability of the modeled double ring square SRR

To summarize, particle swarm optimization (PSO)-based CAD package is demonstrated for design optimization of the building blocks of metamaterial applications, such as square SRR and circular SRR. This user-friendly CAD model is used for the estimation of structural parameters of various configurations of split ring resonators at a desired frequency range. Equivalent circuit analysis in conjunction with PSO makes the CAD model more accurate and computationally efficient.

Another CAD package for path loss prediction in outdoor environment has been developed using *artificial neural network* (ANN) and is described in the Section 9.2.

9.2   Path Loss Prediction in Urban and Rural Environment

Cellular technology has been one of the most rapidly growing fields over the past two decades and has now become an integral part of our day-to-day lives. The first mobile communication system was analog in nature and used frequency division multiplexing for channel access. Over time, these systems were replaced with digital systems incorporating frequency division multiple access and time division duplexing. Network companies implemented cellular architecture and this shift increased the capacity of the channels. However, despite these advances, the biggest challenge in implementing a mobile communication network has always been path loss due to the environment and fading.

9.2.1 Overview

Signal propagation in a cellular network is influenced by the terrain around it. Basic cellular models establish that factors like distance between the transmitter and receiver, height of the transmitting and receiving antenna from the ground, transmitted power and carrier frequency [Rappaport, 1996] affect the received signal. Using these factors, the path loss in the system can be determined by employing the Friis transmission formula.

In addition to the free-space path loss, signals are also prone to multiple reflections resulting in fading and other phenomenon. Generally, the propagation of these waves is discussed in terms of reflection, diffraction and scattering. The EM wave is reflected off multiple surfaces in its vicinity.

Buildings, etc., have the ability to diffract and scatter these waves. As a result, the receiver receives multiple copies of the same signal over a finite interval of time. These signals interfere and result in fading. When the interference is destructive in nature, the signal strength decreases rapidly and the receiver is said to be in a location of deep fade [Rappaport, 1996]. If the receiver is moving, then it may pass through multiple fade regions resulting in rapidly fluctuating levels of received signal strength. In extreme cases, this might even lead to the call being dropped.

In order to overcome these problems, different methods to increase the capacity of networks, for example microcells, have been introduced. Microcells are smaller than conventional cells and are being extensively used in personal communication systems (PCS). Naturally, these microcells are also susceptible to the issue of fading. Therefore, in order to design an efficient and reliable communication network, engineers need to develop methods to predict the fading in the cell. While mathematical formulations for the same exist, corresponding computations are extremely cumbersome.

It is evident from open literature that artificial neural networks have been used as a tool for microwave modelling, simulation and optimization [Devabhaktuni et al., 2001]. In this section, artificial neural networks are used to predict the path loss in outdoor microcells. Following this, the same algorithm is used to develop a JavaScript based CAD model. The developed CAD model yields path loss prediction without going through the process of rigorous mathematical calculations.

9.2.2 Propagation model and path loss prediction

The first step for developing an ANN based path loss prediction technique is the generation of data that is accurate and practical. In order to achieve this, it is essential to use a propagation model that is heavily based on experimental data. As a result, the COST231 HATA model [Rappaport, 1996] is selected in order to obtain the path loss for different values input factors, viz. antenna heights, transmission power, carrier frequency, and distance between transmitter and receiver.

The COST231 HATA model used here can predict the path loss for carrier frequencies ranging between 1500–2000 MHz. The model can accommodate a transmission distance from 1 km to 20 km. The height of the transmitting antenna can be any value between 30 m and 200 m, while that of the mobile station antenna can range from 1 m to 10 m.

9.2.3 CAD package using ANN

As mentioned earlier, ANN has emerged as an effective optimization tool for microwave designs. This development model carries out three steps in the course of optimization: *data generation*, *neural network training* and *testing*. The neural network, therefore, has the ability to learn from its data and the environment, which is the biggest reason for its increasing popularity. The inputs are passed through a neural network that is trained to achieve a certain target. A mapping between the inputs and the targets is performed and optimized weights and bias are determined. This mapping is accomplished using a multilayer perception trained in backpropagation mode [Raida *et al.*, 2003]. The ANN algorithm has been discussed in detail in Chapter 2.

9.2.3.1 Generation of data

Initially, the neural network has no knowledge about path loss determination or the factors influencing it. Therefore, an accurate path loss prediction model, the COST231 HATA model, is used to generate this data. The inputs taken are height of the base station antenna, height of the mobile station antenna, the distance between the two antennas, transmitted power, and carrier frequency. Data is generated for 450 combinations of these inputs, with different data sets for urban and rural environments. Further, the data is divided into three sets: *for training, testing* and *validation*. Once this is done, the data is imported into the ANN program for training the neural network to this data set.

9.2.3.2 Training of the ANN

Once the data is generated, the multilayer perception is trained in the backpropagation mode. The usage of the backpropagation mode enables easy mapping between the input and the required targets. Weight upgradation is performed using

$$w_{ith}^{k+1} = w_{ith}^{k} - \eta \frac{dE^k}{dw_{ith}} \tag{9.8}$$

where, η is the learning rate and E^k is the mean square error at the kth instant. The neural network is hence trained to achieve the targets by learning from input–output samples.

9.2.3.3 Testing

The training of the neural network is stopped once a desired accuracy in the target is achieved. The weights and bias are then extracted and tested using a testing program. Should the test program give the same results as the neural network, it may be concluded that the model has been fully developed. The model can then be used without any further training.

Different neural network models are developed for metropolitan and rural areas with different training parameters in each model. These training parameters affect the efficiency of training.

Table 9.4 Training parameters for artificial neural network

Parameters	Value
Number of input neurons	4
Number of output neurons	1

Number of hidden layers	2
Number of 1st hidden layer neurons	60
Number of 2nd hidden layer neurons	20
Training time (on a 66 MHz P4)	25 min.

9.2.4 CAD model

The optimized weights and bias for the neural network model is obtained by following the procedure given in Section 9.2.3. The inputs and outputs are set and the neural network structure is depicted in the form of weights and bias at each of its synaptic point. The inputs are multiplied and the bias added layer by layer. This CAD model is developed using JavaScript. The CAD model was run and the outputs were obtained in a very short time frame.

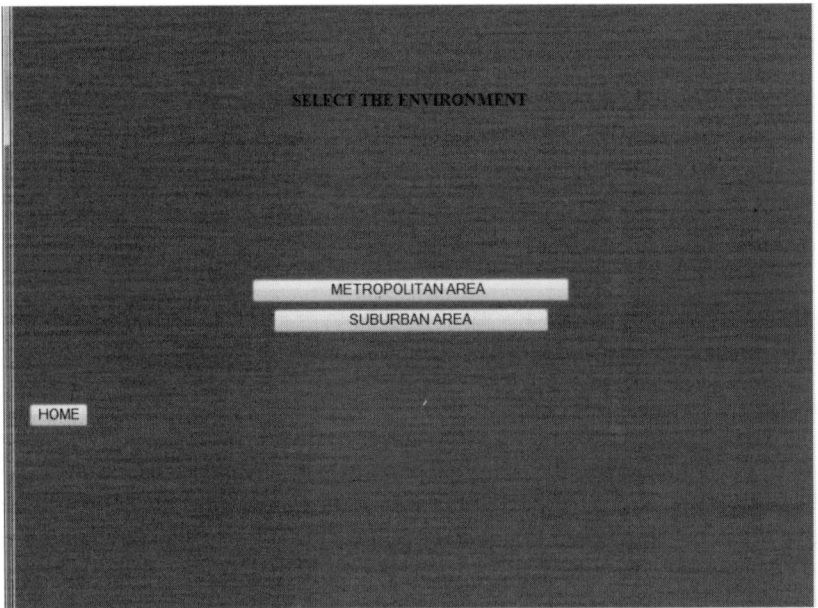

Fig 9.14   Home screen of CAD package

In the CAD package, the user is allowed to select the environment for which path loss must be computed. The user is then re-directed to the screen given in Fig. 9.15 or Fig. 9.16 depending on the choice. When the user selects metropolitan area, the screen in Fig 9.15 appears, the CAD package computed the path loss for metropolitan environment for a base station antenna of height 35 m and mobile station antenna of height 3 m placed at a distance of 12 km from each other. For an operating frequency of 1800 MHz, the path loss for this communication system was found to be 173.767 dB.

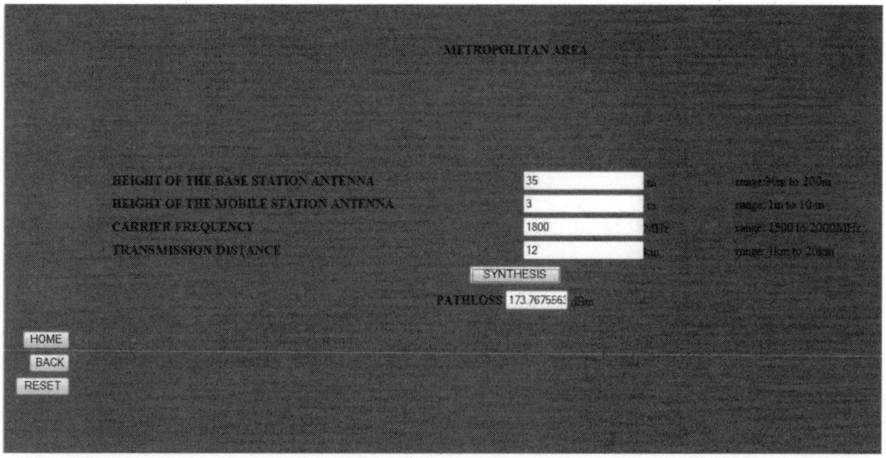

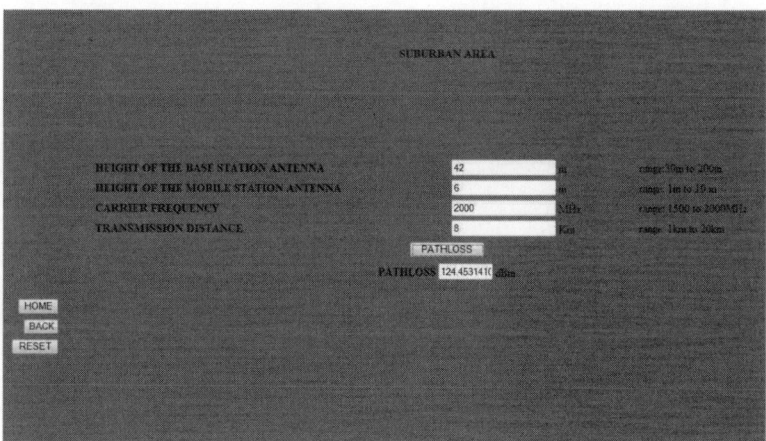

Fig 9.15   Screen for path loss prediction in metropolitan environment

Similarly, for a suburban environment with base station height of 42 m, mobile station height of 6 m, distance between the antennas of 8 m and operating frequency of 2000 MHz, the path loss was found to be 124.45 dB.

Fig 9.16   Screen for path loss prediction in suburban environment

9.2.5 Results and discussion

The performance of the artificial neural network is studied by comparing it with results obtained from an empirical model. The COST231 model is chosen for this study. The inputs for this comparative study are taken to be the same for both techniques. The height of the base station antenna is taken to be 30 m, while that of the mobile station was taken to be 1 m. The two antennas are assumed to be placed at distances ranging 1–20 km. A carrier frequency of 1500

MHz is chosen. The study was conducted separately for metropolitan and rural environments. The results for this comparison are shown in Figs. 9.16 and 9.17.

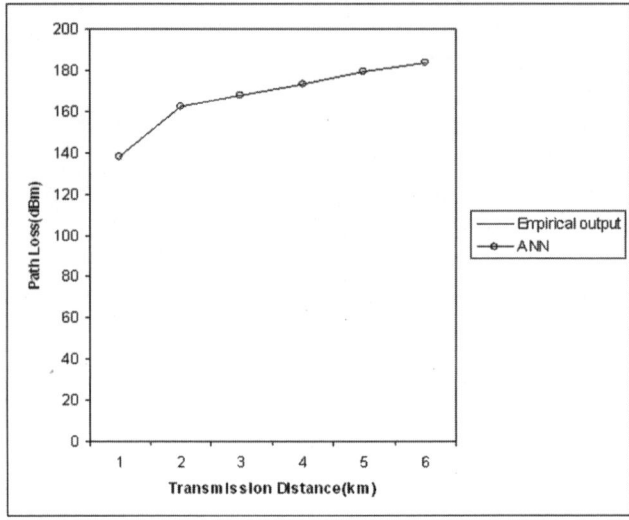

Fig 9.17 Comparative study of ANN output and empirical output for outdoor microcells in a metropolitan area

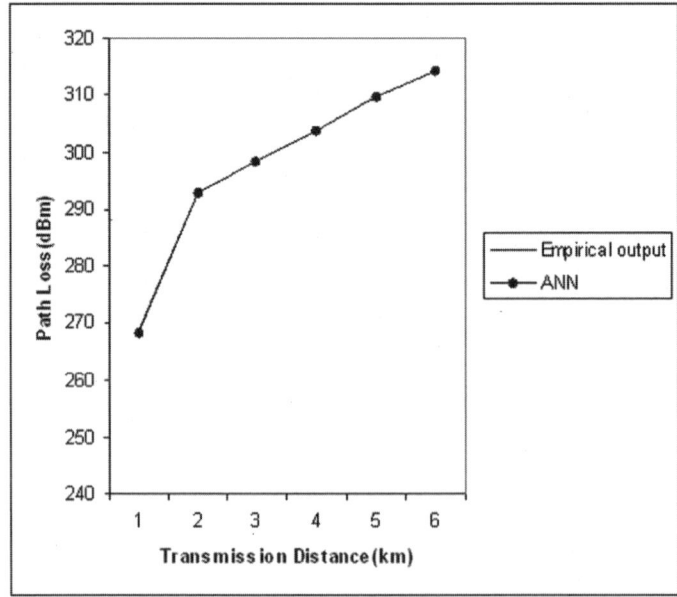

Fig 9.18 Comparative study of ANN output and empirical output for outdoor microcells in a suburban area

The results obtained from the ANN model are found to be in excellent match with the results obtained from theoretical models. This establishes the effectiveness of ANN as a tool for determination of path loss in various cellular environments.

9.3 Summary

The chapter includes two CAD packages developed using PSO and ANN. The first section proposes a CAD model for metamaterial design. A PSO-based user-friendly CAD Package is developed for design optimization of metamaterial split ring resonator configurations. This CAD model uses PSO in conjunction with equivalent circuit analysis method for estimation of structural parameters of various configurations of split ring resonators at a desired frequency range. The configurations considered are single ring, double ring, and triple ring square SRR, besides a circular SRR. A comparison of computational time and accuracy between two different optimization techniques such as genetic algorithm (GA) and particle swarm optimization (PSO) is given in this chapter. The developed CAD package is also validated through simulation software. A graphical user interface has been provided to the EM designers, which makes the CAD package more user-friendly.

In the second section, an ANN approach has been used to design a JavaScript based CAD package. This CAD package reduces the need for cumbersome mathematical calculations. At the same time, the values of path loss obtained from this CAD package are in agreement with those calculated using the COST231 empirical model. This validates the ANN-based model and at the same time demonstrates the effectiveness of the same in the determination of path loss in wireless communication.

References

Baena, J. D., J. Bonache, F. Martin, R. M. Silero, F. Falcone, T. Lopetagi, M. A. G. Laso, J. Garcia-Garcia, I. Gil, M. F. Portilo, and M. Sorolla, "Equivalent-circuit models for split ring resonators and complementary split ring resonators coupled to planar transmission lines," *IEEE Transactions on Microwave Theory and Techniques,* vol. 53, pp. 1451–1461, Apr. 2005.

Bilotti, F., A. Toscano, and L. Vegni, "Design of spiral and multiple split-ring resonators for the realization of miniaturized metamaterial samples," *IEEE Transactions on Antennas and Propagation,* vol. 55, pp. 2258–2267, 2007a.

Bilotti, F., A. Toscano, L. Vegni, K. Aydin, K. B. Alici, and E. Ozbay "Equivalent-circuit models for the design of metamaterials based on artificial magnetic inclusions," *IEEE Transactions on Microwave Theory and Techniques,* vol. 55, pp. 2865–2873, Dec. 2007b.

Choudhury, B., Sangeetha M., and R.M. Jha, "Particle swarm optimization for multiband metamaterial fractal antenna," *Journal of Optimization,* vol. 2013, DOI: 10.1155/2013/989135, pp. 1–8, 2013a.

Choudhury, B., Thiruveni B., P. V. Reddy, and R.M. Jha "Soft computing techniques for terahertz metamaterial RAM design for biomedical applications," *Computers, Materials & Continua,* vol. 37, no. 3, pp. 135–146, 2013b.

Devabhaktuni, V. K., M. C. E. Yagoub, Y. Fang, J. Xu, and Q. Zhang, "Neural networks for microwave modeling: Model development issues and nonlinear modeling techniques," *International Journal of RF and Microwave Computer Aided Engineering*, vol. 11, pp. 4–21, 2001.

Fan, J. W., C. H. Liang, and X. W. Dai, "Design of cross-coupled dual–band filter with equal-length split-ring resonators," *Proceedings of Progress in Electromagnetics Research*, PIER 75, pp. 285–293, 2007.

Goudos, S. K., and J. N. Sahalos, "Microwave absorber optimal design using multi-objective particle swarm optimization," *Microwave and Optical Technology Letters*, vol. 48, pp. 1553–1558, Aug. 2006.

Hata, M. "Empirical formula for propagation loss in land mobile radio services," *IEEE Transactions on Vehicular Technology*, vol. 29, pp. 317–325, 1980.

Haykins, S., *Neural Networks: A Comprehensive Foundation*, Prentice Hall International, NJ, ISBN: 9780139083853, 842p., 1999.

Ivsic, B., T. Komljenovic, and Z. Sipus, "Time and frequency domain analysis of uniaxial multilayer cylinders used for invisible cloak realization," *Proceedings of conference on ICECom*, pp. 1–5, Sep. 2010.

Jin, N. and Y. R. Samii, "Particle swarm optimization of miniaturized quadrature reflection phase structure for low-profile antenna applications," *Proceedings of IEEE Antennas and Propagation Society International Symposium*, vol. 2, pp. 255–258, Jul. 2005.

Kennedy, J. and R. Eberhart, "Particle swarm optimization," *Proceedings of IEEE International Conference on Neural Networks (ICNN'95)*, vol. IV, pp.1942–1948, 1995.

Kwon, H., Z. Bayraktar, D. H. Werner, U. K. Chettiar, A. V. Kildishev, and V. M. Shalaev, "Nature-based optimization of 2D negative-index metamaterials," *Proceedings of IEEE Antennas and Propagation Society International Symposium*, pp. 1589–1592, Jun. 2007.

Mishra, R. K. and A. Patnaik, "Neural network-based CAD model for the design of square patch antenna," *IEEE Transactions on Antennas and Propagation*, vol. 46, pp. 1890–1891, Dec. 1998.

Mukherjee, S., R. Chaudhuri, and C. Saha, "Square split ring resonator: A new approach in estimation of resonant frequency," *Proceedings of 12th Symposium on Antennas and Propagation*, APSYM 2010, pp. 146–150, Dec., 2010.

Oishi, J., K. Asakura , and T. Watanabe, "A communication model for inter-vehicle communication simulation systems based on properties of urban areas," *International Journal of Computer Science and Network Security*, vol. 6, no.10, pp. 213–219, Oct. 2006.

Pradeep, A., S. Mridula, and P. Mohanan, "Design of an edge-coupled dual-ring split ring resonator," *IEEE Antennas and Propagation Magazine*, vol. 53, no. 4, pp. 45–54, Aug. 2011.

Raida, Z., "Modeling EM structures in neural network toolbox of Matlab," *IEEE Antennas and Propagation Magazine*, vol. 44, no. 6, pp. 46–67, 2002.

Rappaport, T., *Wireless Communications: Principles and Practice*, Prentice-Hall, NJ, ISBN: 9788131728826, 707p., 1996.

Robinson, J. and Y. R. Samii, "Particle swarm optimization in electromagnetics," *IEEE Transactions on Antennas and Propagation.*, vol. 52, no. 12, pp. 397–407, 2004.

Sotiroudis, S. P., K. Siakavara, and J. N. Sahalos, "A neural network approach to the prediction of the propagation path-loss for mobile communications systems in urban environments" *Proceeding of Progress in Electromagnetic Research Symposiym,* pp. 162–166, Aug. 2007.

Author Index

Subject Index